Industrial robots and cobots

Everything you need to know about your future co-worker

ISBN 978-83-952513-1-3: Paperback
ISBN 978-83-952513-0-6: eBook

Edited by Magdalena Myślińska
Translation and editing by Sonia Bulanda and Piotr Tokarek
Interior illustrations design by Katarzyna Bąk
Book cover design by Artur Chasikowski
DTP by Maciej ZUCH Mazurek

„(..)Robots are not people. They are mechanically much better than we are, they have an amazing ability to understand things, but they don't have a soul."

Karel Čapek
„R.U.R. (Rossum's Universal Robots)"

1

INTRODUCTION

It was quite a common sight to find horse-drawn carriages distributing blocks of ice through the streets of New York from the mid-19th till the 20th century. Their presence during the summertime was just as common as the presence of the distinctive yellow cabs of the present day. Prior to the invention of freezers, ice was only available seasonally. Its availability during hot days was possible solely thanks to the stock supply accumulated in winter. The ice blocks were obtained from frozen rivers and lakes and then were kept in storage warehouses insulated with sawdust. From there, they were transported to food stores, restaurants and private houses.

Fig. 1: Delivery of ice was not a man-only type of job. The picture presents young girls performing this physically demanding task.

It is estimated that at the end of the 19th century there were approximately 90,000 job positions related to ice trade and delivery in the United States only [1]. Demand for ice in industry and common households was huge – during the peak period, in the 1890's, Norway alone exported over 1,000,000 tons of ice per year. Recipients of the perishable goods were scattered across the world – ice was transported to India, Australia, China and Cuba. It made it possible to store food products for a longer period of time, and it was used medically as well (for example, to help break a fever).

Frozen water, which was available all throughout the year, allowed cities to grow. Prior to that, all cities needed designated agricultural space within their borders in order to provide their citizens with food. Basic refrigeration through the usage of ice blocks extended the shelf life of food products, making it possible to transport them from distant agricultural regions of the country. Fishermen remained at sea for longer, because of the possibility to refrigerate caught fish, and the people on land got used to cold beverages and ice cream during hot days.

The beginning of 20th century is often described as the golden age of ice providers in the United States. The automotive industry started to develop dynamically, and the ice formerly transported by horses started to be distributed by cars. Its transportation did not require high qualifications; it was a task suitable for uneducated people, as well as the immigrants who had difficulties speaking the language. In order to have the goods available by morning, their workday began at 4 AM and lasted until the very evening. It is also important to mention that many delivery workers worked seven days a week.

Despite being very profitable and thriving, the business started to melt down drastically in the 1930's, when refrigerators were first introduced to the US market and were made commonly available for households. We had to wait somewhat longer for that to come to pass in Europe – as they became common in the 50's of the previous century.

The upcoming changes were foreshadowed by industrial installations for ice production that were created much earlier than household devices. They became available in the middle of the 19th century and made the food industry independent from nature's quirks. At the very beginning, the mechanised manufacture was very expensive; machines were unreliable and the ice clarity was visibly worse than that of natural products. Despite being distrustful at first, clients started to appreciate artificial ice over time. In 1914, New York produced 26 million tons of frozen water – it was the first time that this result exceeded the amount of ice acquired from rivers and lakes (24 million tons).

Nowadays, there is no trace of the ice processing and distribution industry that provided jobs for thousands of people. Industrial refrigeration practically killed off ice trading and any activities related to it. However, it has provided other jobs in its place. The job market began to require experts in the fields of design, manufacture, and maintenance of the refrigeration equipment; to perform more responsible, less tiring, and better paying jobs. One could say that technological progress is the reason why thousands of people lost their jobs. However, I think that it is closer to the truth to say that it also offered a chance for a better life for the people who were ambitious and prepared for changes. Those who decided to take up the new jobs were not the only beneficiaries of this change. Lives of the whole society improved drastically thanks to the accessibility of fresh food (often transported from exotic places) and better sanitary conditions.

People responsible for ice delivery began to be slowly replaced by engineers, salesmen, and repairmen who, instead of carrying ice blocks weighing from 10 to 50 kilograms, had non-strenuous and better paying jobs. Trade press and periodicals quickly adapted to the new reality. *The Ice Trade Journal* was rebranded into *Refrigerating World*. Today, it is estimable that in 2009 there were over 113 million refrigerators in American households. 23% of American citizens own two or more such devices. The world has changed.

Ice production is just an example of many fields of industry that have disappeared from the market. The world changes every day. To depict this, try to think about the process of selling used cars, booking a hotel, paying the bills, or sending a postcard from holidays. Now compare these to the stories that your parents have told you. The technological progress and automation which has changed various aspects of our lives has us question the shape of our future.

Today's job market changes because of robotisation, especially in manufacturing companies. When a robot replaces six people, it does not mean that they will have a guaranteed job related to robotics after proper training. Saying that robotisation creates the same or similar number of jobs is naive. Nevertheless, we cannot ignore the positive impact it has on the society as whole.

Nowadays, in the 21st century, 90% of the Burkina Faso population continues to perform strenuous agricultural jobs. There are still many places on Earth where the majority of the society works all day long, seven days a week, regardless of whether or not they want to – simply just to survive. In the United States, the percentage of people working in the agricultural industry is equal to 2%. Not so long ago, in 1870, 50% of the society was involved in agriculture. What exactly has changed over those 150 years?

Since the late 19th century, American agriculture has been revolutionised – fertilizers, tractors, combines, and the modern equipment that allowed farmers to efficiently cultivate large areas was introduced to the market. Mechanisation is responsible, both directly and indirectly, for the situation where many people became unnecessary for agricultural works in the field. For example, when horses became redundant, then naturally, the services of blacksmiths started to lose popularity as well. It does not mean that all blacksmith workshops suddenly transformed into agricultural equipment manufacturers. In a situation like this, the statistics would still indicate 50% of the population working in the field of food production, albeit less directly than before. A part of the blacksmith workshops switched to produce agricultural equipment, but the majority of them went out of business, and their employees have started taking up other professions.

Agricultural mechanisation was also responsible for decreasing food prices, so it was no longer required to work the whole day in order to satisfy basic needs. People had more free time at their disposal and they could spend it on education or perform other jobs that helped them pay the bills: some people took ballet classes while others become nurses or architects. Mechanisation enhanced the collective lives of our entire society by decreasing the required manpower which could have been used for more abstract tasks. Technological progress significantly increased the standard of life in well-developed European, North American and Asian countries. Children could finally attend schools instead of working, and cherish their childhoods. Older people could rest during their retirement, and parents were able to devote more time to raising children and spending more quality time together.

It is the result of robotisation as well. Instead of performing hard and tiring jobs, people can work less and spend more time developing their hobbies. Robotisation creates many jobs directly related to designing, operating and servicing, but most importantly, it gives people time for the activities they were too busy to perform before.

The marginalisation of the importance of numerous professions is a process that is still in progress. Simple and monotonous tasks are often handed over to robots, and the employees that formerly performed them are assigned to other jobs. It creates a demand for experts in other fields. The job market needs people with the knowledge on robot servicing and programming, as well as supervision over and the optimisation of the manufacturing process. The job requires more responsibility, is less dangerous and physically demanding, and is better paid.

The current demand for robot programmers is significantly higher than the supply. European experts help with commissioning production lines all over the world – mostly thanks to the automotive industry. Brazil, China, United States, RSA, Mexico – those are the countries where European countries have their contracts signed, and many of them are struggling with an insufficient number of hands to work. Robotics is an industry that utilises very expensive technologies; therefore, potential employers would like to see people with at least a basic knowledge of robotics and control systems during job interviews. There are very few books that explain the rules of industrial robotics in an approachable way – this makes it very difficult for new employees to enter the job market.

I would like this book to be of assistance for people who are considering a robotics engineering career, students of automation and robotics faculties, and people who are considering changing their field of expertise and who want to work in this growing industry (automation specialists, design engineers). It is also a source of helpful information for people working in factories that invest in industrial robots. This book contains all the information required to become a robot operator. I would like robotisation to be perceived as an opportunity of job promotion, and not as a potential threat to existing job positions.

I learn about the complexity of robotics every day while executing projects for the automotive industry. State-of-the-art robots are usually tested during the car manufacturing process, and then it is made available for other industries as well. In this book, I would like to share my professional experience gained while working for companies such as BMW, Mercedes, Volvo, Porsche, and Lamborghini in Europe and in the United States.

Industrial robotics has been developing very dynamically over the past few years, mainly thanks to collaborative robots, self-propelled platforms and virtual reality. The first fifty years of the industrial robots' development has been devoted to the competition between manufacturers to adjust their products in the best possible way to adhere to the requirements of the companies that manufacture hundreds and thousands of products daily. As a result, nowadays we have access to devices with high efficiency, speed, and lifting capacity. Robots manufacturers have recently started to pay close attention to the needs of small companies. This led to designing devices completely different from those in the past catalogues. Presently, we can purchase robots that respond to touch and can be controlled and programmed when they are "held by the hand" in learning their task. In the near future, we can expect interactions based on voice commands. All of this is supported by advanced security systems that allow people and robots to work side by side. Changes in manipulators' and safety systems' structure will soon cause the popularity of robots to grow even more dynamically – we are much closer to the moment when robots will be present in our homes and offices.

This book is not a robot programming tutorial. It is impossible to write just one book on robot programming, because there is no universal command language for manipulators of different manufacturers. Every single one of them developed a unique language applicable exclusively for its products. It might appear that the next obstacle to learn robot programming is much harder to surpass, because people think that learning robot programming requires an actual robot. Writing applications for computers or smartphones is a skill that can be learned by practically anyone (because computers and modern phones are easily accessible), but access to the industrial manipulator is much more of an abstract matter. Even if you are lucky enough to work for a company that possesses such devices, or if your school's laboratory is equipped with these, such devices likely do not stand there idle anyway.

Fortunately, there are robot simulators that can accurately simulate real behaviour of robots. It is not necessary to own a real-life physical robot in order to learn about programming or rules of robotics; its virtual counterpart is sufficient enough.

The market offers access to 3D simulation programs that allow its users to build complete robotised production lines and check how they operate. Those applications are usually equipped with a user-friendly interface, embedded tutorials, and the option to install it on a private, non-dedicated computer – all of these features attract users. Currently, all of the big players among robot manufacturers and companies such as Siemens Industry Software or Dassault Systèmes offer programming and simulation systems. Those programs are a very good learning method, allowing one to get information on the syntax of languages used in robotics, and also on work practices applicable in the industry.

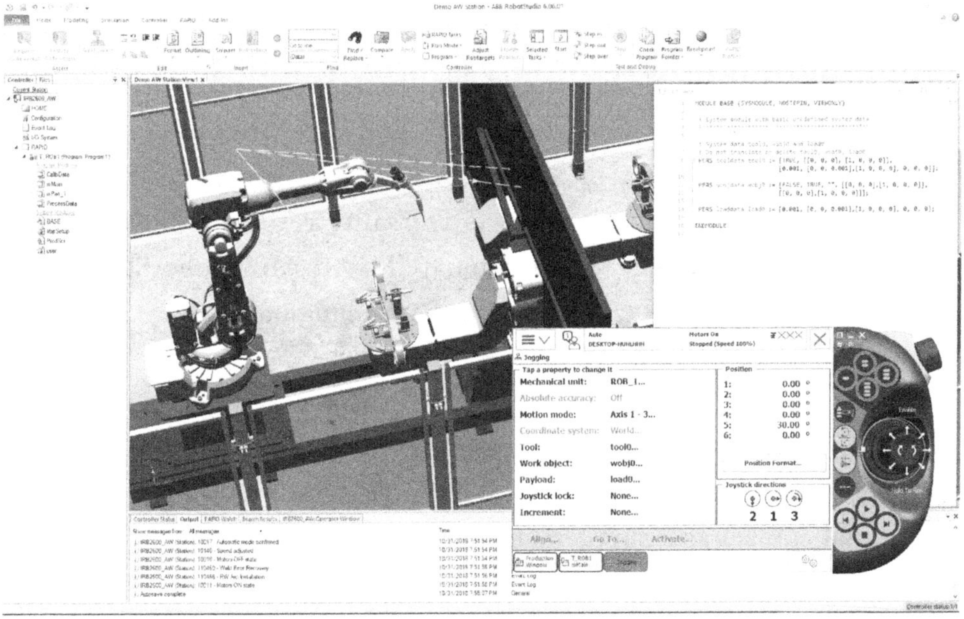

Fig. 2: Virtual model of robotised workstation in a tool for robot programming and simulation.

I would personally recommend the ABB RobotStudio program, which can be downloaded from the producer's website (roboticsbook.com/RobotStudio/). It is equipped with a complete library of components and a virtual control panel that is identical to the one that you will see in real-life robots. The application is a fully functional 30-day trial version, which is just enough time to learn about the examples provided in this book. The program can still be used after the trial period expires, but some of its features will be inactive.

If you think that observing robots on a computer screen is not enough and you would like to have physical contact with the device – there is always an option to purchase a mini-sized manipulators. Kickstarter portal often shows campaigns with the goal of building a small programmable robot. Some of those projects succeeded. You can find a list of such devices on the following website: roboticsbook.com/small-robots/ Prices start from approximately $ 500.

The second program that I would like to recommend is Tecnomatix RobotExpert by Siemens Industry Software (roboticsbook.com/RobotExpert/). It allows you to simultaneously create programs for robots of various manufacturers. Its universality results in fewer brand-specific options available when compared with ABB RobotStudio, but it allows you to learn various programming languages and is available in a trial version as well.

Make sure to visit the following website while reading this book – roboticsbook.com/bonus/. You will find some additional materials there, that can act as a supplement to this book.

2

HISTORY OF ROBOTICS

Dreams about a mechanical helper relieving humans from their duties are as old as humanity itself. To trace the history of robotics, we could go back to ancient times and search for the seed of the idea that led to the creation of the first robot. Considering the fact that this book is about industrial robotics, we will go only go back by roughly about one hundred years, to the year 1920 when in his play, *R.U.R.* (*Rossumovi univerzální roboti*), Karel Čapek used the term "robot" for the first time. He described a fictional industry manufacturing workers like machines. Therefore, we can consider this fragment to be the first literary definition of a robot:

> *"Miss Glory, robots are not people. They are mechanically much better than we are, they have an amazing ability to understand things, but they don't have a soul."*[1]

In Karel Čapek's play, robots are human-shaped creations manufactured in large numbers, and their task is to relieve humans from their strenuous tasks. Present-day robots are nothing like the robots described by the Czech playwright.

This is not an isolated case in regards to the history of technology. We are surrounded by objects that had a different initial concept compared to that of present day. A good example are airplanes, which do not flap their wings as birds do.

Apart from his famous quote on the cars' colour, Henry Ford has also accurately summarised his clients' expectations:

> *"If I had asked people what they wanted, they would have said faster horses."*

[1] Original text: „Drahá slečno Gloryová, Roboti nejsou lidé. Jsou mechanicky dokonalejší než my, mají úžasnou rozumovou inteligenci, ale nemají duši [2]

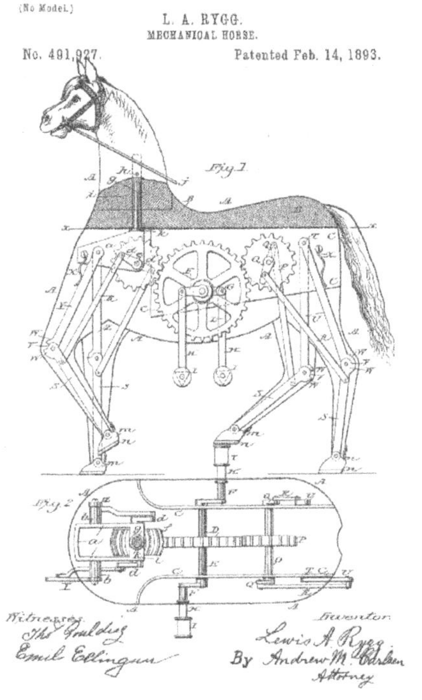

Fig. 3: United States Patent and Trademark Office has received a patent claim for a mechanical horse.

This is exactly what the beginnings of an automotive industry looked like. The first person to come up with the idea of a vehicle that would utilise mechanical legs was Thomas Brunton [3]. This idea has also been executed by an English man – David Gordon. In 1824, he built a vehicle that moved in accordance with the same rules as the horse's limbs did. This "car" was equipped with 6 steel legs powered by a steam engine. They had special feet that were pushing the device from the ground and putting it in motion. The feet were equipped with springs and levers imitating sinews that allowed them to adjust their length accordingly to surface irregularities, and walk smoothly on a bumpy road. Despite being complicated, the structure was working well, although it moved forward very slowly. Additionally, this type of drive was partially responsible for destroying the road surface it was moving on, causing Gordon to discontinue developing his construct and stop further experiments.

Ultimately, the machines have replaced animal traction, and Karl Benz proved that a car does not have to be a mechanical horse when he built the tricycle. Robotics also needed someone who would prove that creating a man of steel is not the right way to go.

2.1. Development of robotics around the world

Less than forty years have passed from the first use of the term "robot," to the moment the first industrial specimen was constructed. An American constructor, George Devol, had started work on its construction in the early 1950s. The patent concerning a device that can perform repetitive tasks (called *Programmed Article Transfer*) was submitted in 1954, and granted 7 years later.

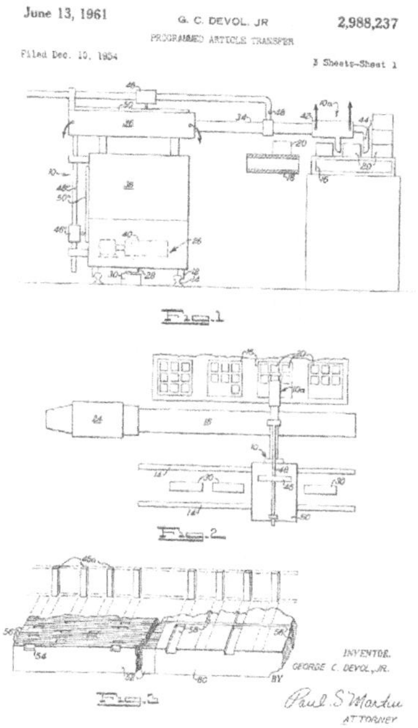

Fig. 4: Patent for the first device that was an industrial robot.

The first prototype of a robot was created in 1959 by George Devol and Joseph Engelberger. The fact that both men knew one another is pure coincidence as they met in 1956 at a cocktail party in Connecticut. When Engelberger heard about Devol's patent during their talk (which he commented on with the following words: "That sounds like a robot to me"), he managed to convince Devol to start a collaboration. Both of them wanted to design a device that would satisfy the real needs of a market. Therefore, the prototype construction was preceded by a series of visits in manufacturers (mainly automotive, e.g. General Motors, Ford, Chrysler). They named their company Unimation, from the phrase *universal automation*, and their first robot was called "Unimate 1900".

Engelberger was lucky enough to meet very interesting people in his life. He completed his Master's Degree in Physics at the Columbia University in New York in 1949 – the same university where Isaac Asimov completed his doctorate in Chemistry just a year earlier. Asimov was the creator of the term "robotics" and the three laws of robotics he described in 1942 in the *Runaround* book. Engelberger was familiar with his works, from which he had borrowed the word "robot". This word is probably the only Slavic word related to technological development that is being used internationally. Apart from some exceptions (German: *roboter*, Finnish: *robotti*, Spanish: *roboto*, Portuguese: *robo*), the whole world uses the word "robot".

The Unimate 1900 robot was an almost 2-ton giant, and his program was recorded on magnetic drums. Programming was performed with the utilisation of joint coordinates, and the whole structure moved thanks to hydraulic drives. Positioning accuracy was approximately 2.5 mm. Short uptime of the first robots was an issue. The MTBF rate (*Mean Time Between Failures*) was equal to barely 400 hours [4]. Modern robots can proudly present the same rate at the level of 70,000 hours.

The Unimation company founded by Devol and Engelberger quickly acquired the first customer for their product. In the spring of 1961, the first Unimate started to work at the Trensted manufacture in Trenton (New Jersey). The facility belonged to General Motors and produced handles for windows and doors, gear shift knobs, and other interior furnishings. Unimate has found its use in the process of die-casting. The robot moved hot casts to the cooling basin, and handed them over to workers after they had harden for further industrial treatment. The units moved were not heavy – they weighed from 13 to 18 kilograms, but working with these was very monotonous, and the workpieces themselves were red-hot. A moment of distraction could have resulted in a severe accident. Therefore, it was a perfect test ground for a new device such as a robot.

Despite initial concerns, Unimate turned out to be a huge success – it has performed burdensome and dangerous tasks that were previously handled by men working three shifts. Having worked for over 100,000 hours, the robot was handed over to the Smithsonian Institution. Before the purchase, General Motors corporation estimated that the life expectancy of this robot would only be 18 months.

The first robot in Europe was installed in 1967 in a Swedish factory Metallverken in Uppsland Väsby, where it was responsible for executing monotonous tasks of moving products from one place to another [5].

Engelberger was promoting Unimate in the media, e.g. in *The Tonight Show*, where a robot played golf, poured beer and conducted an orchestra. Despite the problems with providing stable pressure in the hydraulic system, the TV show turned out to be successful and the audience was very enthusiastic about it. Workers employed in factories were far more sceptical about robotisation and the introduction of robots. They worried that the machines will substantially decrease the number jobs.

The General Motors Corporation, satisfied with the results of Unimate, installed the first spot welding robots in 1969, in the assembly factory in Lordstown, Ohio. The factory received 26 new devices that performed a total number of 520 spot welds on the body of a newly introduced Chevrolet Vega (code name: XP-887). Thanks to automation, only about 10% of 3900 welds were performed manually. At the same time, the remaining factories that were not using robots performed only from 20% to 40% of spot welds automatically.

In order to increase the efficiency of welding in traditional manufacturers, multi-spot welding machines were used that could make many welds at once. Those devices had limited possibilities; they were built with a specific element of a car's body in mind, and if any part of the structure changed the whole machine had to be rebuilt. Robotisation of the production process allowed the factory in Ohio to speed up their production to 100 cars per hour – two times more than any other car factory at that time. New cars were leaving the assembly line every 36 seconds. The Dodge Main factory at that time produced 64 cars per hour, and the Ford Mahwah facility's efficiency was equal to 52 units [6].

From a technical perspective, the factory in Lordstown was a huge success that presented the technological capabilities of the GM corporation. At the time when Chevrolet Vega's production started, it was the most automated car factory in the world. However, working conditions and the quality of cars produced there were not something to be proud of. A very short work cycle caused the workers to perform their duties carelessly and a heavy workload accounted for a frequent turnover of employees.

At first, the factory was divided into two departments: Chevrolet and Fisher Body. This type of organisational structure caused the surplus of employees, resulting in a situation where the assembly line was supported by far more people that was actually required for production. This surplus made it easy for the employees to quit and rest. Troubles arrived in 1971 when both these departments were merged into one: General Motors Assembly Division (GMAD). After the merger, the cost needed to be optimised so that "the surplus" of employees was fired, and, at the same time, the production speed increased. The management board wanted to make a quick return on the money invested in the factory construction in 1966, and that resulted in more pressure on the production rate.

The General Motors Corporation considered this factory to be the factory of the future. Maintaining a high rate of assembly was not a problem for robots and machines.. The very assembly that distinguished GM quickly became the nightmare of its employees. Due to a short production cycle, the assembly tasks were divided into numerous easy tasks. The whole factory was organised in a manner that limited unnecessary movements of the employees, which resulted in repeating the same task thousands of times, over and over during one shift. This lasted for weeks, which turned to months. The employees were treated as if they were machines; there was no buffer time for a human factor, such as dropping a part or a tool. The activities that had to be performed by a worker in one minute beforehand now needed to be performed within 30 seconds. There was no time for breaks, because there was nobody that could substitute for them. Horrible working conditions quickly found themselves reflected in a car's quality – Chevrolet Vega was known for its frequent failures.

Being totally consumed by the possibilities offered by robots, without changing the employees' work environment, quickly showed its aftereffects. The infamous strike against scandalous working conditions took place on May 3rd, 1972. It lasted 3 weeks and cost the General Motors Corporation 150 million dollars. This strike started the discussion on working conditions in (a more and more automated) automotive industry in the United States.

In 1969 the market has seen the first industrial robot intended for painting (Trallfa TR-2000). Trallfa was a small engineering company located in Stavanger, Norway that was producing agricultural equipment. The company had a lot of troubles finding employees to spray-paint wheelbarrows due to hazardous working conditions. Ole Molaug, a young engineer, decided to challenge this problem, and in 1966, he designed the first robot intended for painting. Ole needed one more year to improve his construction and prepare it for industrial use. The device was equipped with hydraulic drives, and the program was written on a magnetic tape. The robot could mimic the movements of an experienced sprayer that were shown to it during the learning process. Apart from its capability to learn continuous movements, this device also had another advantage over Unimate products – it was much cheaper.

Towards the end of the 1960's, Japan was growing at a rate of 12% GDP a year and it experienced a significant labour shortage. It comes as no surprise that the American industrial robots got proper attention. 1969 was the year when an enterprise agreement between Unimate and Kawasaki Heavy Industries was signed. The agreement concerned production and purchase of manipulators on the Asian market. This cooperation resulted in the creation of Kawasaki-Unimate 2000 in the same year. It was the first industrial robot manufactured in Japan. The collaboration between Kawasaki and Unimate lasted 15 years and it ended with producing an overall number of 2400 devices.

Over the course of the next two decades, robotics developed significantly. New concepts from the field of kinematics and control systems started to emerge. Numerous solutions that were developed in 1970s and 1980s are still being used today.

In 1971, KUKA GmbH installed the first production line in the Mercedes factory in Sindelfingen, Europe and equipped it with hydraulic robots. Two years later, the same German company started producing its own robots and in 1973 it introduced a robot called "Famulus," the first manipulator in the world equipped with six electric motors.

In 1974, Kawasaki built its first robot – based on Unimate's solutions – developed for the purposes of arc welding, which is used for the production of motorcycle frames. In the same year, Fujitsu-Fanuc, a renowned Japanese manufacturer, instituted its first robot designed for numerical control machines.

1974 was also the year when ASEA Irb 6 made its debut in Europe. It was the first industrial robot equipped with electric motors and a microprocessor controller (with an 8008 chip delivered by a company that was relatively unknown at the time – Intel). The anthropomorphic design of the manipulator resembled human arm. The robot had 5 axes of movement and a payload of 6 kg. The S1 controller was the first to use an Intel 8 bit microprocessor and its memory capacity was 16 kB. The controller had 16 digital I/O and was programmed with the use of 12 keys and a four digit LED display. Irb 6 construction works were performed in 1972-1973 and the final product was introduced in October of 1973. Two years later, the ASEA IRb 60 with a payload capacity of 60 kg was developed. This robot was an answer to the requirements of an automotive industry which needed heavier payloads. The manipulator was first delivered to a Saab factory in Sweden where it welded car bodies.

The first half of the 1970's was the time when companies manufacturing robots saw an increase in their sales. These constructions have reached a level of a matured technology and they already had a wide audience of potential recipients who were interested in them. In 1973, there were 3000 robots were in circulation worldwide and 30% of them were produced by Unimation.

In 1974 Hitachi developed 'Mr. AROS' – an arc welding robot. It was equipped with a microprocessor and a gap sensor that allows it to correct its path by detecting an accurate location of the welded parts.

In 1978 Hiroshi Makimo from University of Yamanashi develops a SCARA-type robot. SCARA is an acronym for *Selective Compliant Assembly Robot Arm.* Thanks to its specific construction, the arm of the manipulator was rigid in the Z-axis and pliable in the XY-axes, which is extremely useful in mounting activities. This feature allows the robot to connect the parts that are not perfectly aligned with one another. Sankyo Seiki launched the production of a new type of robot in 1981.

In 1978, a PUMA robot was introduced. It was developed by Unimation (which got its design along with the acquisition of the Vicarm company in 1977) in collaboration with GM. Having noticed that 95% of parts for passenger cars weigh somewhere between 1.5 and 2.5 kg, General Motors Corporation requested the construction of a compact robot with small payload capacity. Even though the first industrial robot was developed in order to help out humans in the hardest working conditions, PUMA was designed for the purpose of replacing an employee in tasks that are too easy to be done manually. PUMA manipulators were licensed and manufactured by many companies, including the Finnish Nokia, which produced about 1500 devices in the 1980's. Apart from the production of mobile phones, the previous world leader also manufactured and designed robots (Nokia NS-16, NRS-15).

Aside from the PUMA robot design, Unimate also made use of the programming language that Vicarm invented – VAL (*Variable Assembly Language*). It was one of the first languages used for programming robots.

In 1979, Nachi company presented its first electromotor-driven spot welding robots. In the welding process, there were splinters, impurities and a very strong electromagnetic interference present. For this reason, one had to use a proper shielding to ensure that the controller and robot drives could work reliably.

In 1985, KUKA develops a robot with new Z-shaped robot arms. The robots developed up to this date had parallelogram-based mechanisms, which limited the freedom of movements. It also caused the movement of axis 2 to be dependent from axis 3 and vice versa. This really complicated the programming process and it limited the available workspace. The parallelogram-based mechanisms are still used today, but, in most cases, they are used in palletising robots. A new approach to the construction has resulted in building more compact devices.

The development of robotics led to the establishment of the International Federation of Robotics (IFR) in 1987. Its first task was to develop criteria that allowed for the creation of uniform definitions in the field of robotics. Lack of common terminology and a straightforward classification made it difficult to obtain reliable data on the number and types of robots installed. In some countries these devices were classified in a different way, which had a negative impact on the development of the industry as a whole. It was difficult to forecast future demand or the upcoming trends in robotics. JIRA (*Japan Industrial Robot*

Association), which was founded in 1971, was the first organisation in the world that dealt with robotics.

The first Delta robot was sold in 1992 by a Swiss company called Demaurex. Its construction was based on the idea developed by Professor Reymond Clavel from EPFL (*l'École Polytechnique Fédérale de Lausanne*) in 1985. An exigency for the invention of this new type of robot was the need to automate chocolate pralines packaging. Since sweets are rather light, the first idea was to build a robot with very light arms designed to promote quick movement.

In 1988, Unimation went bankrupt and it was acquired by Staübli. The bankruptcy of the first robot manufacturer was mainly caused by their belief in hydraulic motors' efficiency, whereas the customers purchased more electric-driven robots. Robots started performing more accurate processes (i.a. arc welding) and Unimate products were not as precise and fast as their counterparts. Unimate licensed its patented innovations to other companies, which brought it a stable revenue, but also resulted in inviting more competitors to enter the market. The licensees have put a great effort into the development of Unimate's solutions, which was also the reason why they manufactured better products over time.

In 1994, Motoman presented its innovative MRC controller. It was the first control system that provided a synchronised control of up to two robots. MRC made it possible to edit robot programs on a computer and to control up to 21 axes at the same time.

KUKA was also eager to introduce innovations in its control systems. 1996 was the year when the first controller with a PC – KRC1 architecture was presented. Its control panel was equipped with a 6-D manipulator used for controlling the robot and a user-friendly window interface. Three years later, the same manufacturer integrated into its robots the functions that enable online access to diagnostics and monitoring of the robots' work. In 2000, ABB decided to update its controller which resulted in the processor Pentium MMX becoming the heart of a new S4C+ unit.

It was 2006 when Comau, an Italian company, introduced the first Wireless Teach Pendant (WiTP). It was the first time when one could control a robot from a distance greater than the length of a controller cable.

The KUKA KR1000 Titan was presented in 2007. It was the first robot that had all its axes powered by electric servomotors and with a payload capacity of 1,000 kg. Not much later Fanuc introduced a robot with a payload capacity of 1,350 kg (M-2000iA/1350 model), and in 2015 one could purchase the M-2000iA/2300 model that was able to lift up to 2.3 tons.

The most recent years in the industry were devoted to manufacturers competing to win in the field of greater energy efficiency and more compact size while maintaining the same payload capacity and increasing the operation speed. Most main manufacturers nowadays offer robots that can be controlled with touch-operated controllers that look like a tablet.

Another challenge that this industry brings is the construction of safe and easy-to-use robots for small enterprises. What distinguishes them from their bigger counterparts is their user-friendliness. Nowadays, every significant manufacturer has such a device on offer.

Automotive industry is one of the most developed economic sectors in terms of automation. This industry was the first to use robots and it is also the largest recipient of industrial robots at present times. If we compare it to the general industry we can see significant differences in the application of robots. However, this situation changes every year which makes the differences become less substantial. Manipulators are becoming more and more universal. They perform numerous tasks that we did not expect them to do a few years ago. Robots are even present in amusement parks, they make pizza, or they become camera operators in film studios.

This is just a summary of all the changes that have taken place within the last sixty years and it surely does not cover all of them. It is also worth noting that enormous progress has been made in industrial networks, sensors, machine vision systems and robots' tooling.

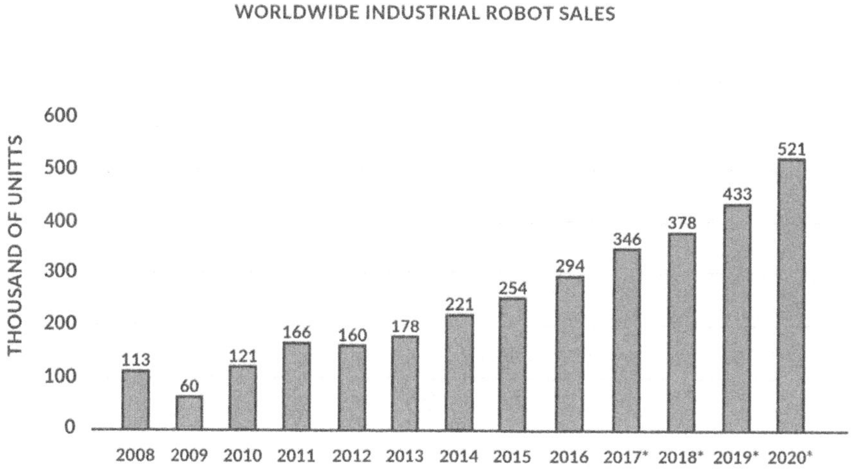

*Fig. 5: Global industrial robot sales (*a forecast) Source: IFR World Robotics 2017.*

It is even more important to note that the market is continuously changing – new sectors and applications are introduced, and there is a systematic increase in the global sales of industrial robots.

3

STRUCTURE OF INDUSTRIAL ROBOTS

When Joseph Engelberger, one of the fathers of robotics, was asked what an industrial robot is, he responded:

> *"I can't define a robot, but I know one when I see one."*

This statement was too vague and not formal enough to become an official definition, which led to the formulation of a more precise definition in ISO 10218:1 norm (*Robots and robotics devices*)[7]:

> *Industrial robot: automatically controlled, reprogrammable multipurpose manipulator, programmable in three or more axes, which may be either fixed in place or mobile for use in industrial automation applications.*

Since the first robot was introduced to the market, numerous structure forms of manipulators were developed. Many of these solutions were functional for quite a while until newer models were introduced which gained a lot of attraction and phased out their predecessors. At present, when we say "industrial robot", we think about a serial structure manipulator with six degrees of freedom that uses rotational kinematic pairs (each drive with an electric motor). With every new generation of these devices, they become more universal and facilitate the performance of tasks that had to be executed previously only by devices of a highly specialised structure.

Abundance of universal robots with electric servomotors is clearly noticeable when skimming through an old manipulator catalogues. Forty years ago, when one wanted to buy a manipulator that could work in an explosive environment, they had to choose a model that was driven by compressed air (no risk of sparks). Only hydraulic robots could lift heavy payloads and manipulators with electric drives were only used whenever precision and ease of control was crucial. Nowadays, it is highly unlikely to see any hydraulic or pneumatic robots in the market. The progress in material development, CAE simulations, and the construction of transmission and servomotors has made it possible to construct a robot with lifting capacity exceeding 2 tons (Fanuc M-2000iA/2300). Additional sealings, brushless electric motors and spark-proof protection eased the construction of units in accordance with ATEX directive that are able to work in the environment with an explosive atmosphere. In the 1980s, module robots (the ones that can be put together from standard components) were thought to be the future of robotics, whereas nowadays they constitute only a niche exception.

The number of solutions that were developed within fifty years from the introduction of the first robot allows for the creation of complex divisions and a classification. In this book we will focus on a few main structure forms – the ones that are presently used, gain popularity, or whose position will not be endangered in the next few years.

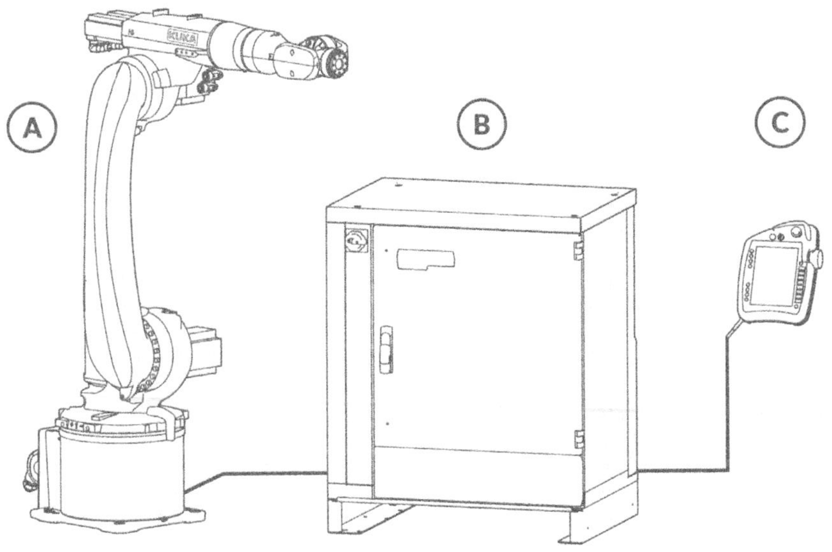

Fig. 6: Complete robot: a) mechanical unit b) control cabinet c) teach pendant.

An industrial robot is made of three main elements:

- mechanical unit and its motors
- control cabinet

- teach pendant, also referred to as: operator's panel, control panel or control console.

Mechanical unit defines robot's range, its precision, lifting capacity and speed. Robot controller is responsible for controlling the mechanical unit's drives, communicating through built-in interfaces and managing safety systems. The operator's panel (teach pendant), which is attached to it is akin to a tablet that is enclosed in a handy case. Thanks to the teach pendant we can send commands, edit programs and keep track of robot's work.

With every generation of their products, the manufacturers try to introduce higher lifting capacity, reduced weight, lower cost, easier maintenance procedures and unify the parts with their other products' components. A great quality progress can be observed not only in the construction of a mechanical unit but also of a robot controller. It is already a thing of the past to store programs on magnetic drums or by using jump wires. Even though we are experiencing a continuous progress, we can still see devices that have been working for almost 20 years (which is oftentimes what manufacturers take great pride in). Robots are designed in a way to work and be reliable for many years. That is also the reason why 3½-inch floppy discs may still be a part of robot programmer's daily life. It should come as no surprise then that a person working on a cutting-edge laptop may have a USB external floppy drive plugged in; it is by no means an archivist from a computer science museum, it is a robot programmer.

Development of the controllers is similar to the progress happening in consumer electronics. This dynamic advancement may result in purchase of used units that could cause some unexpected problems. The robot is a subdivision of a larger whole and it is necessary for it to be able to work with other elements of the station. A large number of used robots come from automotive plants, where high-tech, reliable, and very expensive solutions are being used. Second-hand devices that has been in service for many years are not considered modern anymore. Control cabinet may limit the possibility of using high-tech peripherals (machine vision systems, tools, elements of the safety system), because it cannot communicate with them. Additionally, finding modern peripherals supporting the old communication type is either impossible or very expensive.

In case of using a second-hand device, one should take into consideration not only the higher likelihood of failure, but also other problems related to it. All failures may require finding spare parts for the control cabinet, which may not always be in stock and their prices may be much higher than for the spare parts of the supported solutions. Another problem with old controllers is that they do not always work well with high-capacity storage devices. You can try for yourself how much time it takes to find a CompactFlash memory card with a capacity of 4 MB will take you (yes, 4 megabytes).

Of course, this should not be a reason to discredit used robots, but simply to raise awareness of the problems that one may encounter. Presently, there are many enterprises that specialise in renovation and sales of used manipulators. Using the services of renowned companies may be translated into savings when buying a robot. However, it is very important to be aware of the services they offer, i.e. the availability of their maintenance services and spare parts for the robot model one is willing to buy. Robot manufacturers guarantee that spare parts for their products will be available for at least 10 years since the discontinuation of their production.

It would not be taken as an advantage if choosing an older robot would mean using old peripherals due to its lack of compatibility with the newer ones. In some cases, this problem could be eliminated by using an old manipulator and a more up-to-date controller that supports present communication standards. Manufacturers offer used manipulators that have been completely refurbished and restore their full functionality (*ABB Certified Robot Refurbishment, Yaskawa Remanufacturing Program, KUKA Used Robot*). Distributors sell display robots that were in operation for an overall of several dozens of hours during fairs and exhibitions. These are the devices that are basically new and have full warranty, but they are offered for a lower price, just like ex-demonstrator cars at car dealerships.

There are also new enterprises on the market that offer industrial manipulator rental (RaaS – Robot as a Service). If a robot is to be used for many years at the assembly line, then it is advisable to get a loan and buy a new one or lease it, but in case one needs a robot for a short-period of time or for testing, then rental proves to be a very convenient solution. Making your company's stand at the fair more appealing can serve as an example here. Robot at the fair draws attention but it is not economical to buy such an expensive device just so that it operates for a few weeks a year.

Another important thing one needs to consider when buying a robot: within the meaning of the EU's Machinery Directive (2006/42/EC) a robot (regardless of whether it is used or new) is an incomplete machine and it does not bear a CE marking. To obtain a CE marking, it is necessary for a machine to have an appropriate machine guarding and safety system. It is important to remember this in case one wants to install one on their own at a factory. A robot integrator provides you with the complete package, so if you purchase the whole workstation, an integrator should guarantee meeting all standards needed.

3.1. Degrees of freedom

The number of degrees of freedom is equal to the total number of independent parameters needed to explicitly define the position of an object in space. For example: a point on a plane has two degrees of freedom (no possibility of rotation, moves along the X or Y axis). When considering a rigid body, freely suspended in a three-dimensional space, we have 6 degrees of freedom. They consist of the following movement parameters: translation (X, Y, Z in Cartesian coordinates) and rotation (rotating about the above-mentioned axes: Rx, Ry, Rz).

A 6-axis robot can move objects with the use of six degrees of freedom:

- three translating motions along the X, Y and Z axes (translatory motion).
- three rotary motions around the Rx, Ry and Rz axes (rotational motion).

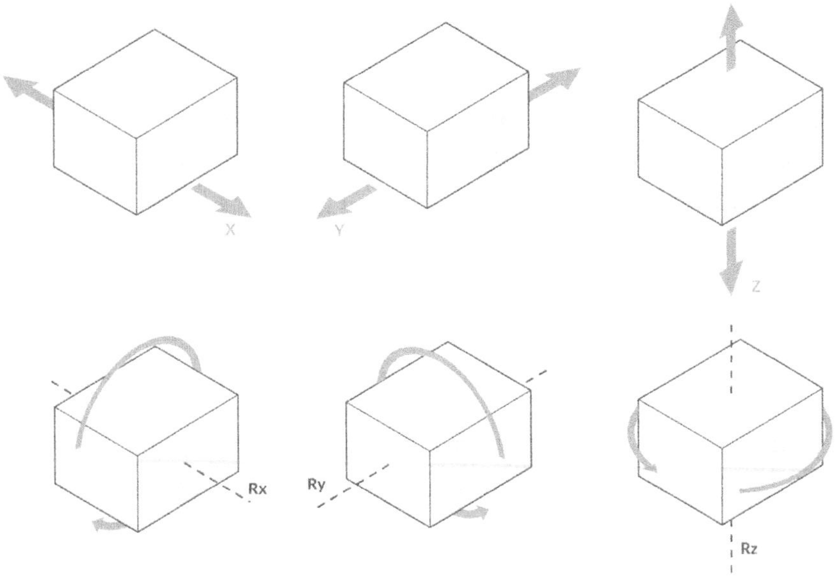

Fig. 7: Degrees of freedom – translation and rotation.

It means that an object can be manipulated as if it was a free body – it can be moved and rotated about X, Y, Z axes of the Cartesian coordinates (of course, this does not mean that a robot will have the full capabilities in every part of the workspace – the more its arms are stretched out, the less manipulation capabilities are available).

Below you can find examples of movements that are available depending on the number of robot's degrees of freedom (DoF):

- One DoF: moving an object along a straight line (e.g. up/down or forward/backward or left/right – one of these).
- Two DoF: moving an object on one plane, e.g. backward/forward and right/left (but without the possibility of changing the orientation and, in this case, without the possibility of moving up/down).
- Three DoF: the ability of moving an object without changing the orientation (up/down, forward/backward, right/left) to any place in space X, Y, Z, as long as the range of robot's arms allows it.
- Four DoF: same as in the previous point plus the possibility of changing the orientation along one axis (e.g. Rx)
- Five DoF: moving an object to any place on Cartesian coordinates (X, Y, Z) and the possibility of changing the orientation along two axes (e.g. Rx, Rz)
- Six DoF: object can be treated as a free body – it can be moved freely in X, Y, Z space and rotate it about three axes: Rx, Ry, Rz
- Seven DoF: compared to a device with six degrees of freedom, it is also possible to change the alignment of robot's arms without the need to move the object, which can be freely positioned in space (X, Y, Z, Rx, Ry, Rz). A human's arm also has more than six degrees of freedom. This allows us to freely position an object held in our hand, and, for example, change the elbow position without the need to change the object's orientation.

A majority of robots have six degrees of freedom, and if they have less than six, then we are probably dealing with a palletizing robot (usually 4 degrees of freedom, which are enough for the robot to freely move objects from one place to another).

3.2. Mechanical unit

The most popular solution is robot with an open-kinematic chain mechanism, where all kinematic pairs permit rotation. Since any chain is made out of links, a kinematic chain can also be defined as a mechanism that is made out of elements (rigid bodies) that are connected to one another. The word "open" means that the last element is not connected to the first one (Fig. 9a).

A connection between two elements is called a kinematic pair. Modern robots use kinematic pairs with one degree of freedom, i.e. those that can be rotated (rotational pairs) or moved (translational pairs) about a single axis (Fig. 8). This solution makes it easier to control the robot, makes it more accurate and reliable. In industrial robots, each kinematic pair is put in motion by a dedicated servomotor. Not all of them are always visible, in some constructions they are shielded and built into the arms of the manipulator.

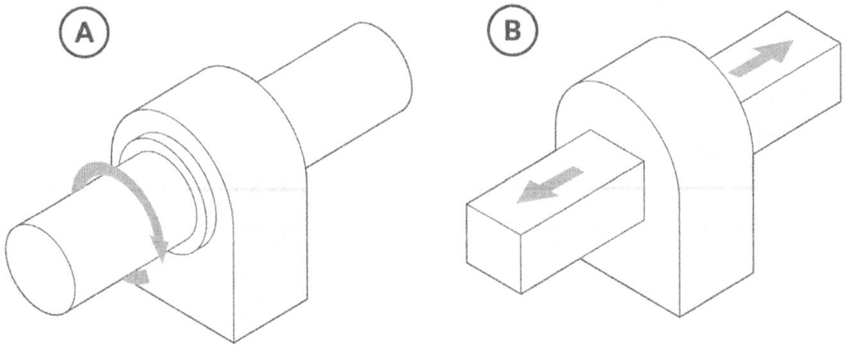

Fig. 8: Kinematic pair with one degree of freedom a) rotational, b) linear.

From the kinematic viewpoint, the manipulators can be divided into serial, parallel and hybrid. In serial robots each consecutive arm is connected to the previous one. Parallel robots, on the other hand, have their links located next to one another and are not directly connected with each other. Hybrid manipulators have distinctive features of the two types described above, as shown in Fig. 9.

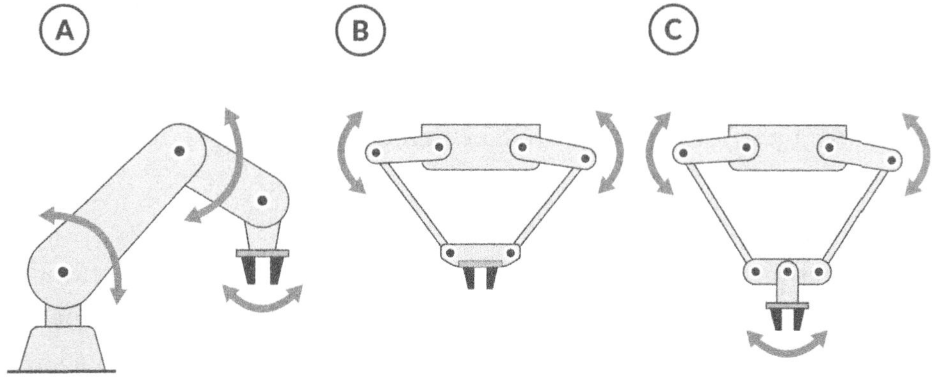

Fig. 9: Manipulators: a) serial, b) parallel, c) hybrid

Robots come in various sizes. The smallest standard industrial manipulator is the size of a standard (A4) sheet of paper (0.26 m range, weight of 4.5 kg – Mecademic Meca500). The biggest robot available is Fanuc M-2000iA/1700L that weighs 12.5 tons and has a range of 4.68 meters.

When examining a real-life robot (Fig. 10) we can easily see its base and consecutive arms. Arms' structure must be able to ensure high rigidness and the joints between them are called axes or joints. The axes are numbered from 1 up, starting from robot's base. The first three joints of robot's mechanism (A1, A2, A3) are called main axes, the last three (A4, A5, A6) make up its wrist.

A standard robot has from four (for palletizing) to six (for other purposes) axes. Though majority of robots have mainly six axes, there are seven axis industrial robots available on the market (e.g. KUKA LBR IIWA, Yaskawa VA1400, Fanuc R-1000iA/120F-7B). Nowadays, the available constructions constitute in robots with small and medium lifting capacity. An additional degree of freedom means that we can operate the robot in a tight working area. This also means that the robot is less rigid. A seven axis robot offers a wider variety of possible arm configurations, which equals to easier programming in very limited work space (roboticsbook.com/7axisrobot/). The presence of one "extra" degree of freedom allows for the position change of a kinematic chain without the need to move and change orientation of a tool mounted on the manipulator, which may be significant for some purposes.

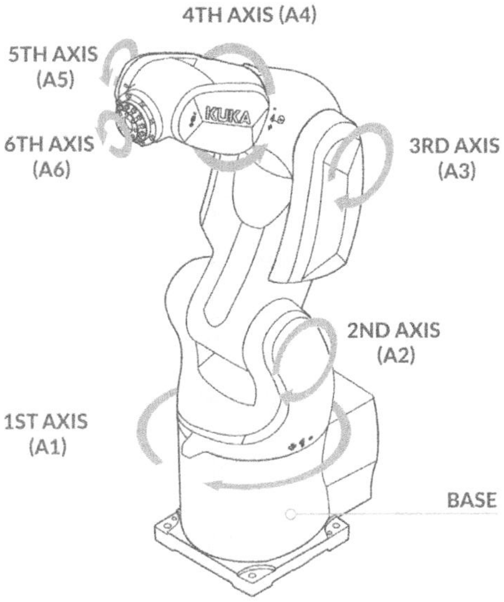

Fig. 10: KUKA KR 3 R540 Industrial robot.

The seven axis robots are much more expensive and their structure is way more complicated than that of a six axis one, that is also why they are not treated as competition for classic constructions with six axes. It is good to know that these exist, but they are just another family of devices and may be used only in very specific cases. It can be easily seen there are very few manufacturers offering these. Lightweight industrial robots with seven axes of movement stand as a base for building more anthropomorphic structures. However, their use for industrial purposes is even more limited; they are generally used for more scientific purposes. DLR Justin or Yaskawa SDA10D are examples of these.

Having seven axis does not mean that the manufacturers have given up on developments. ABB IRB 800 is one of more original manipulators as it has ten axes. This robot is a result of merging two other products: a palletizing manipulator IRB 660 (4 axes) and a six axes IRB 140. It has an impressive range (3.85 m) and can work in hard-to-reach places. It is controlled by a single IRC5 controller. This unique construction can be found, for instance, at ISS Wawrzaszek special vehicles factory in Poland, where it is being used for the purpose of welding fire truck frames. Find out more at roboticsbook.com/irb800/

Robots' manufacturers try to make common subassemblies that they use to construct their products. Quantec, Kuka product family, is composed of 26 types of robots. They are made out of common parts: one type of base, one of two rotational columns and one of two compensation systems. Four types of servomotors and four types of transmission make up these robots. Beside lower production costs, it is also much easier to service them. Higher degree of unification reduces the storage cost of spare parts and it also increases their availability. Robot production is not at all highly automated and most installations are performed manually (roboticsbook.com/robot-factory/).

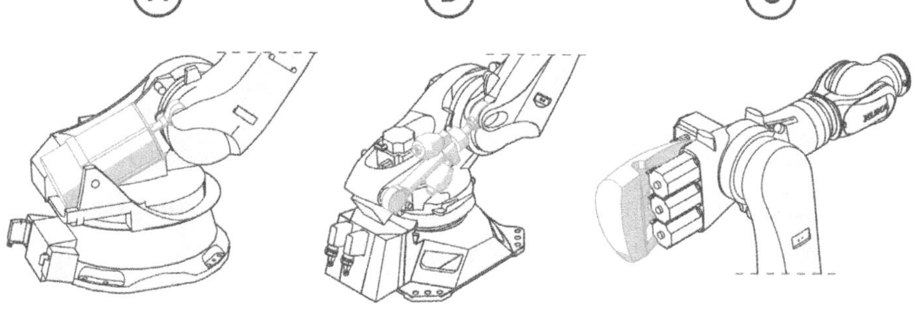

Fig. 11: Compensation system of the second axis of the robot using a) spring energy b) hydro-pneumatic system c) third axis counterweight.

Fig. 12: Warning! Energy stored in counterweight system.

One of the characteristic features of robots with medium and high lifting capacity is a compensation system (balancer) which is usually mounted between the base and first arm of the robot. It is composed of a spring or a hydro-pneumatic system (Fig. 11a and b), which relieves the servomotor at maximum extension of the robot's arms. Please remember that any interference with this system by an untrained person is very dangerous! A compressed spring has a lot of energy and an attempt to open a cylinder where it is placed may lead to life-threatening injuries (as presented on appropriate warnings, Fig. 12).

To relieve the servomotor on axis three, an even easier solution in the form of a counterweight is used (Fig. 11c).

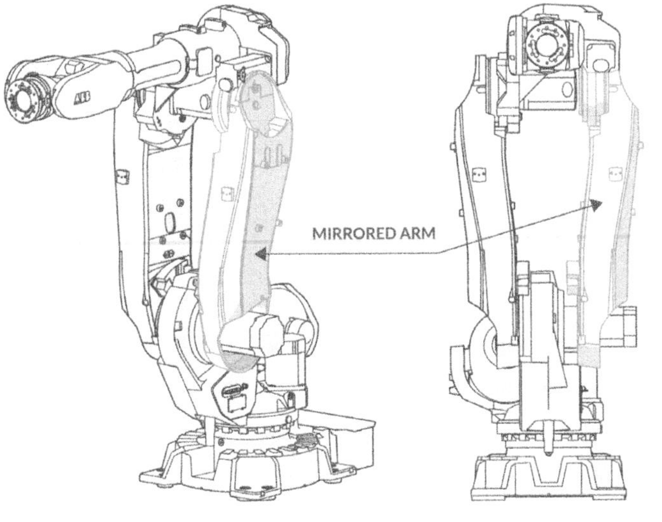

Fig. 13: Virtual model of ABB IRB 7600 robot used for simulation purposes with an outline of a mirrored arm.

Most robots are built in a way that their arms are not symmetrical. It is important to note this while designing robotised station (an example of such device with a visibly asymmetrical structure is Yaskawa ES200RD). In case of robots with high and medium lifting capacity only few devices are symmetrical.

Whenever a production line is designed, where it is certain from the beginning that it will have its mirrored counterpart, special robot models are used for simulation. They have an added "missing" arm (indicated in Fig. 13 by light grey), which can detect the possible future problems related to lack of symmetry of the other side. If no limitations were found during simulation, then it can be assumed with high probability that the second, mirrored station will not experience these obstacles as well. It helps executing design and programming works only once. Then we can transform programs to obtain their mirrored versions in order to save time and money. An example of production line where mirrored robot stations are used is the one where car door are manufactured.

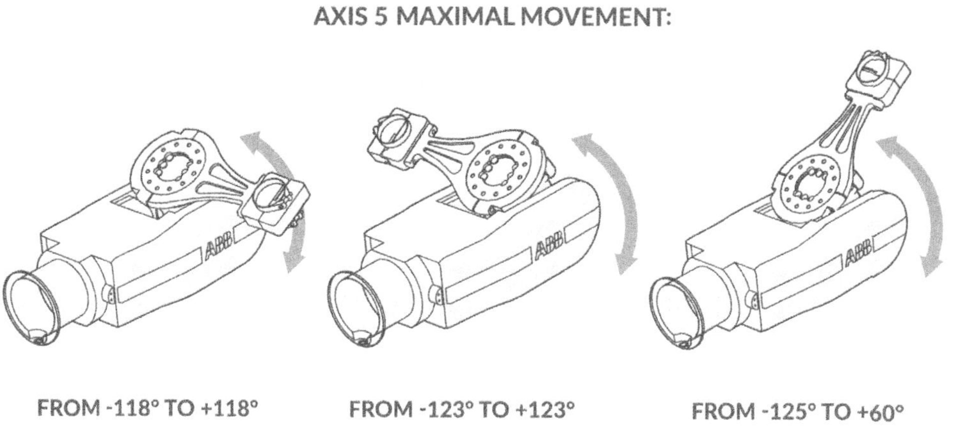

Fig. 14: Examples of limitations caused by lack of wrist symmetry.

Though lack of symmetry of the manipulator's arms is clearly visible, the differences in the wrist construction may not be as evident until we encounter some problems during commissioning. In Fig. 14 we can see how the maximal axis 5 movement range changes depending on the way of mounting the cable bracket on axis 6. The way of mounting the tool may decrease the real motion range available at the end axes compared to what is written in the documentation. That is the reason why using simulation software is a standard nowadays when designing robotised stations. It can help visualise and verify the problems that we may encounter in real life on a virtual model.

When deciding on using a certain robot model, we should consider the shape of a tool flange, as it is the element where tools are mounted. Every manufacturer lists its dimensions in their catalogue and indicates all fastening holes and threading on technical drafting. In most cases, the tools can be mounted on a robot in only one way because of the centering pin. If there is a need to mount a tool at a different angle, additional adapters are used and the tool is mounted to them.

Each axis of the robot has its own drive. Robot servomotors have a limited movement range and none of the axes can turn ad infinitum in one direction only. Exemplary range of motion along with maximal speeds have been indicated in the table (Tab. 1). As you can see, smaller robots have much greater rotational speeds. Range of motion for respective axes depends on the structure of the robot's mechanical unit. While comparing the manipulators of various manufacturers with similar lifting capacity and range, we should be able to easily tell the difference between them (e.g. in the range of motion of the first axis). Each additional range of motion's level makes the robot more versatile, as well as easier to be implemented and programmed.

Tab. 1: Comparison of range of motion and speed of robots with different lifting capacity. [8][9]

Axis	ABB IRB 6700-300/2.70 (300 kg lifting capacity)		ABB IRB 1200-7/0.7 (7 kg lifting capacity)	
	Range of motion	Speeds	Range of motion	Speeds
A1	±170°	100°/s	±170°	288°/s
A2	+85/-65°	88°/s	+135/-100°	240°/s
A3	+70°/-180°	90°/s	+70°/-200°	300°/s
A4	±300°	140°/s	±270°	400°/s
A5	±130°	110°/s	±130°	405°/s
A6	±360°	180°/s	±360°	600°/s

In addition to the mechanical limits (resulting from the existence of physical limitations), the robot can have defined software limits. This way, robot manufacturers limit the range of movement of the last, sixth axis, which in most cases could rotate endlessly, as there are no physical obstacles that would prevent it from doing so. Some manufacturers offer additional paid option of endless rotation for the last axis of the robot.

A robot's controller allows the user to define their own program limits which further reduce the available working space. In practice, this function is used to limit the robot's ability to work in certain areas (e.g. where workers may appear).

The robot moves thanks to the servomotors – electric drive systems with a feedback loop, which allows it to continuously compare the current position of the drive with the set one (closed-loop system). Permanent magnet synchronous machines are basically the standard for today's industrial robotic systems.

Servomotors are characterised by high power density, high efficiency and low inertia. They can work at zero speed – they generate enough torque to keep the robot mechanism in a given position. They can monitor the current rotational speed and angular position of the shaft with high accuracy by using built-in resolvers or encoders (an electro-mechanical device that converts the angular position or motion of a shaft to analog or digital output signals). We can provide ABB IRB 1600 as an example here, where the angular resolution of applied drives is equal to 0.001 degree.

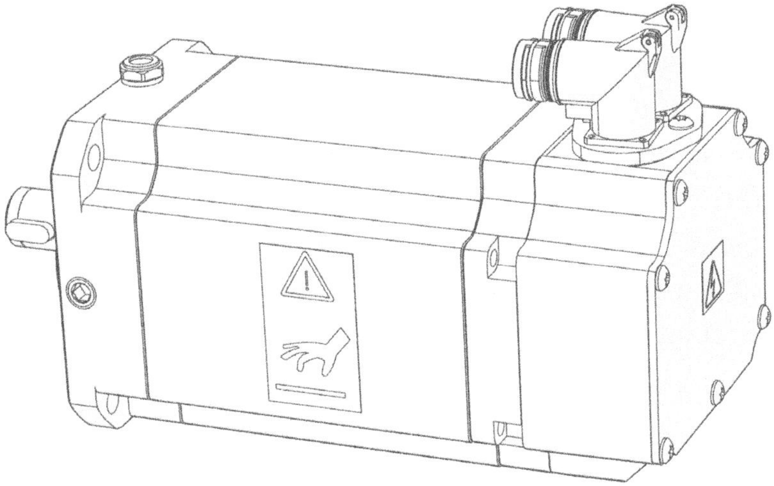

Fig. 15: Servomotor manufactured by ABB.

The run cycle of an industrial robot's servomotor consists of rapid acceleration and equally sudden braking, as well as frequent speed changes. Dynamic operation results in large amounts of heat being generated. A servomotor's casing can heat up to high temperatures that can cause burns if touched – appropriate stickers on the drives warn against this.

Robot's drives are equipped with mechanical brakes that make the robot maintain its position in the event of power failure and during prolonged downtime. They are closed by default. In order to release the brakes, an appropriate control signal should be sent, and if it is not present (e.g. in the event of a power failure), the robot will automatically switch to the braked state. It works exactly the same as in the case of train air brake systems. The brakes are released there as long as compressed air is supplied to them. If the system is unsealed, a drop in pressure causes the train to brake (the system automatically goes into a fail-safe state in the event of a failure).

The robot's brakes have a limited lifespan and require periodic diagnostics. Manufacturers equip their products with diagnostic programs that should be run regularly to verify their performance. In this test, the robot attempts to perform a movement using the maximum torque of the drives while the brakes are active and the entire machine stands still. If the drive's feedback system detects that the manipulator has moved, the test is failed and interrupted, then a warning is displayed. During the test procedure, the robot should have some space around it, so that if the brakes are not able to hold it, there is no damage to the surrounding parts.

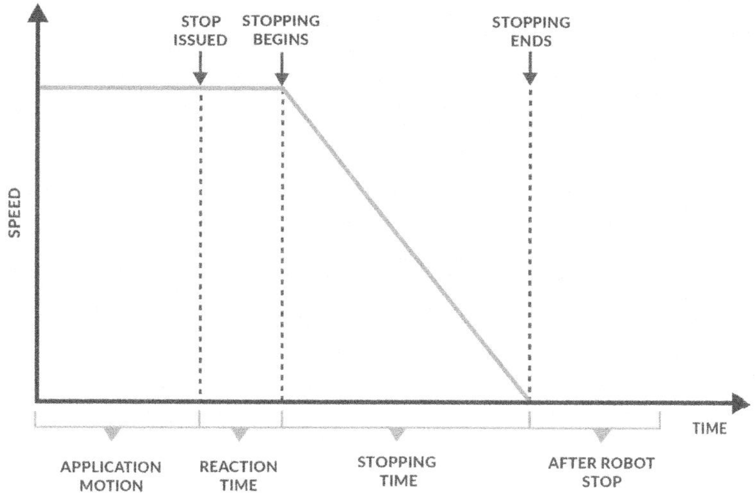

Fig. 16: Robot stoppage time chart.

Although maximum speeds are most desirable from the point of view of reducing the time of tasks execution, they have a downside as well. Bringing a robot with a high payload capacity, and with a maximum speed and load, to an emergency stop takes over 1 second. If you measure the distance travelled on the tool flange of the sixth axis, you will find that it is about three meters! This means that the robot can crush a person on its pathway without any problem, even if it starts to brake before the collision occurs. For more information on safety at work with industrial robots and the associated risks, see chapter 7.

If a severe collision should occur and the robot would not be able to leave on its own, it is possible to release the robot's brakes manually. This can be done by using the buttons on the robot or by using a brake release unit/device, which will give the appropriate signal to release the brakes. Manual release of the brakes is an emergency operation that requires the appropriate protection of the robot! First you need to evaluate if it will not cause more damage than trying to leave the collision manually. The brakes on the axes should be released one at a time, after securing the arms against free fall. It is necessary to control the effect our actions have on the robot arms' behaviour. Only robots with low lifting capacity are light enough to be held by a human; for larger models, appropriate supporting devices should be used.

The robot needs to be calibrated before commissioning and periodically after a certain number of operating hours. Calibration (mastering) is performed in order to achieve high precision robot movement and high positioning accuracy. It consists in defining zero positions for all servomotors, which is required to precisely calculate the robot's position. The mechanical unit has special calibration marks on its arms. Calibration is performed by setting the robot arms positions so that all markers match. While in this position, calibration occurs and the robot controller treats this position of the drives as the "zero" point of reference. Calibration may be required when the robot is disconnected from the power supply for a long time or when the precision of movement falls below the acceptable level – e.g. as a result of a collision. Information on calibration are saved in the memory maintained by the batteries that may get discharged (resulting in robot being decalibrated). Incorrect calibration causes the robot to reach the declared points of the program with an offset.

Fig. 17: Calibration marks for the fifth axis of the robot.

Robot manufacturers require calibration of the device in the following cases:

- before commissioning
- after repairs (e.g. after servomotor or transmission replacement)
- when the mechanical unit of the robot has been moved with the control system switched off
- when the mechanical end-stop is struck at the speed of more than 250 mm/s
- after collision.

The working space of a manipulator can be significantly increased by mounting it on a rail system (also referred to as seventh axis/7th axis/rail). More than one robot can be installed on a single seventh axis. Rail systems are mounted flat on the ground. It is allowed to mount them differently (on the wall or ceiling – e.g. KUKA KL100) only in case of constructions designed for very light robots. Rail systems have a modular design, which makes it easy to adjust their length to the requirements of the process that is being carried out. Control of the position of the manipulator on the seventh axis is executed by the robot's controller. Rail system is visible as another drive (i.e. additional seventh axis of the robot – hence the common name).

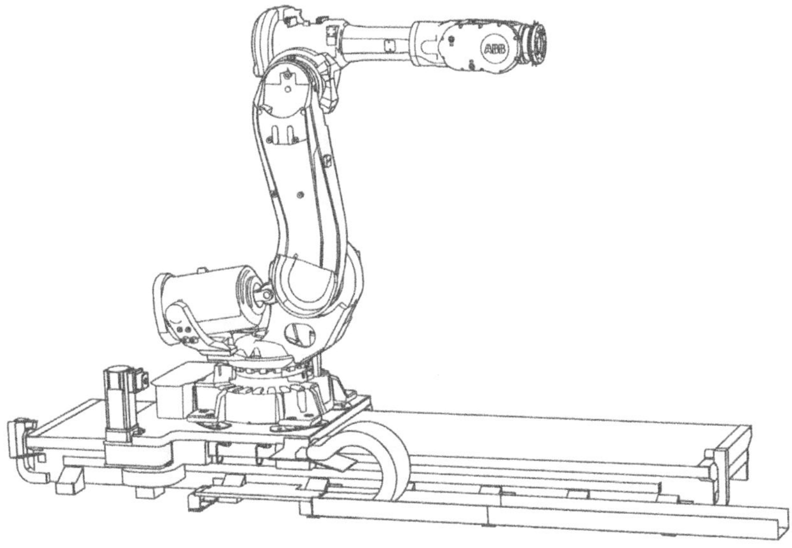

Fig. 18: ABB IRB6640 robot mounted on a linear motion track.

Portals, rotors and positioners additionally make it possible to increase the working space. They are usually used during the arc welding process, allowing the weld to be executed in a flat position, regardless of the course of the weld on the curvature of the part.

Most devices, in addition to being mounted flat on the floor, can also be installed at an angle, on the wall or suspended from the ceiling. This reduces the amount of space occupied by the robot and, as a result, opens the way for further optimisation of the robotised station components' layout. Mounting the robot in different positions and at different heights makes it possible to increase the density of devices working simultaneously, having a positive effect on the cycle time. Information on permitted mounting positions for the given manipulator can be found in the manual. If the robot is not standing flat on the ground, it is necessary to enter information about this fact into the controller, in order for it to be able to properly compensate for the fact that the gravitational force is acting in a different direction than the default one.

Each manufacturer precisely specifies the position in which the robot can be transported. The correct positioning of the manipulator's arms ensures the centre of gravity to be optimally positioned, preventing the manipulator from tipping over on either side during transportation.

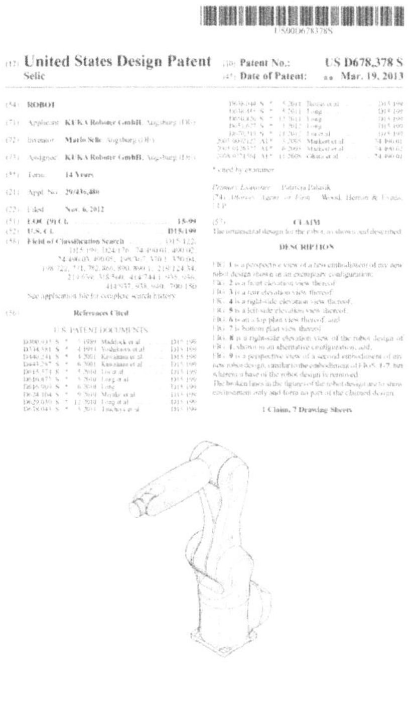
United States Design Patent — Selic
Patent No.: US D678,378 S
Date of Patent: Mar. 19, 2013
ROBOT
CLAIM
DESCRIPTION
1 Claim, 7 Drawing Sheets

Fig. 19: Registered industrial design of the KUKA Agilus robot.

In addition to purely practical advantages, manufacturers are paying more and more attention to the appearance of their robots. Despite the fact that the industrial robotic system is still aimed to be primarily practical, why shouldn't it also be pretty? *The Red Dot Design Award* and the *iF Design Award* are among the two best-known design awards. Robot manufacturers win distinctions in these competitions more and more frequently. In recent years the awards have been granted to:

- *Red Dot Award*: ABB YuMi, KUKA LBR IIWA
- *iF Design Award*: DENSO VS 050/060
- *Good Design Award*: Universal Robots UR5 & UR10.

A robot's design is especially important in case of low lifting capacity robots. While only a small number of steelworks' employees can see the heavy robots intended for the metallurgical industry, collaborating robots are increasingly present in non-industrial applications. They are used in the entertainment industry and present in artistic installations, showrooms and service points. Every year, industrial robots will be increasingly visible in our daily lives.

Manufacturers of industrial robots have followed in the footsteps of consumer electronics manufacturers and more frequently patent designs of their manipulators. When tasks do not require the robot to possess high technical parameters, appearance can be a decisive factor in the purchase. It also makes it possible to differentiate oneself from the competition and to make the customer feel connected to the brand. While interesting appearance has long been required for consumer products, now it is slowly becoming a necessity for industrial products as well.

Each producer paints their products in a colour distinctive to itself. The Yaskawa robot is blue (Munsell 5PB 3.6/12), ABB is white now (RAL 7035/Pantone 428 C), although until recently it was orange (NCS 2070-Y-60R), like KUKA (RAL 2003). Fanuc in turn is painted yellow (Munsell 5Y 8/14). Until recently, Comau was available in the distinctive red colour (RAL 3020), which has been then replaced by a combination of silver and black since 2015.

There is a trend among manufacturers to move away from reflective colours, which used to warn about the danger. Today's robots are intended to work alongside people and calm colours such as white, grey or green are used to signal the technical advancement and friendliness of these devices for operators.

Of course, there is nothing to stop you from ordering the robot in your own, company's colours. That is exactly what the car manufacturer Tesla has done. Their KUKA robots at the Freemont factory come in the same shadc of red as the brand logo.

3.3. Control system and operator's panel

Robot's controller manages the drives, sends and receives signals from the connected devices and the PLC. The capabilities of the controller can be increased by installing extension cards (e.g. to support additional communication protocols). Robot's controller is usually set up as a slave to the PLC, which manages the robot's operation (it sends a start signal, a stop signal and monitors the progress of tasks).

Modern controllers are based on components known from personal computers. Current generation of KUKA control cabinets (KRC4) has been built with the use of Intel Celeron G1820 processor – operating at 2.7GHz frequency. Controllers of many manufacturers are based on Windows and VxWorks real-time system, but proprietary operating systems written by robot manufacturers exists as well. One controller can supervise the operation of more than one robot arm (solutions such as: ABB MultiMove, KUKA Roboteam).

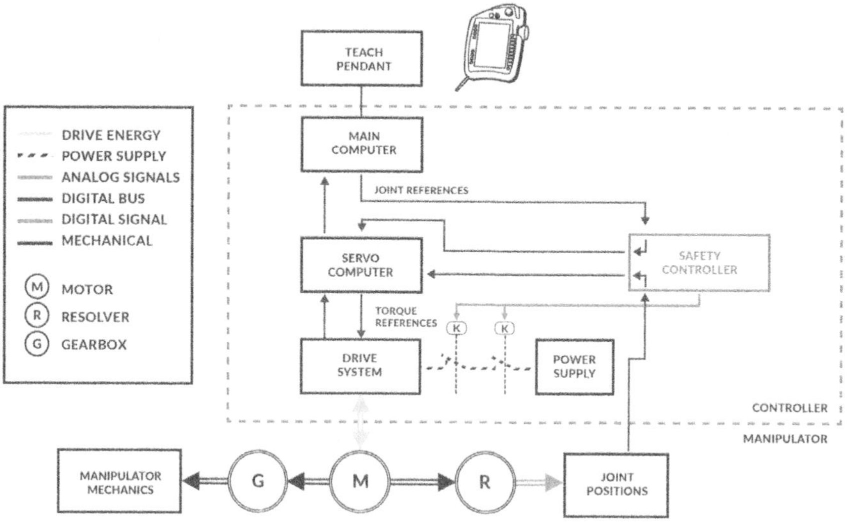

Fig. 20: Diagram of the individual components of the robot controller.

The schematic structure of the robot controller is shown by Fig. 20.

The main computer located in it is responsible for program interpretation, motion planning and calculations of forward kinematics and inverse kinematics. Servo Computer is responsible for controlling the position and speed of the drives as well as the forces applied to the servomotor. The safety controller verifies if the robot's actual movement corresponds to the programmed one, monitoring its speed and position.

Teach pendant is used to control the robot and to view and edit programs. You could say that it's a remote control for the robot: it's a small tablet that can be held comfortably in the palm of your hand. The robot is controlled with a variety of possible solutions. Some manufacturers use only buttons (Fanuc, Comau, Kawasaki), others try to make it more practical by adding a 6D manipulator (KUKA), and some of them decide to add a joystick (ABB).

Fig. 21: Teach pendant manufactured by KUKA.

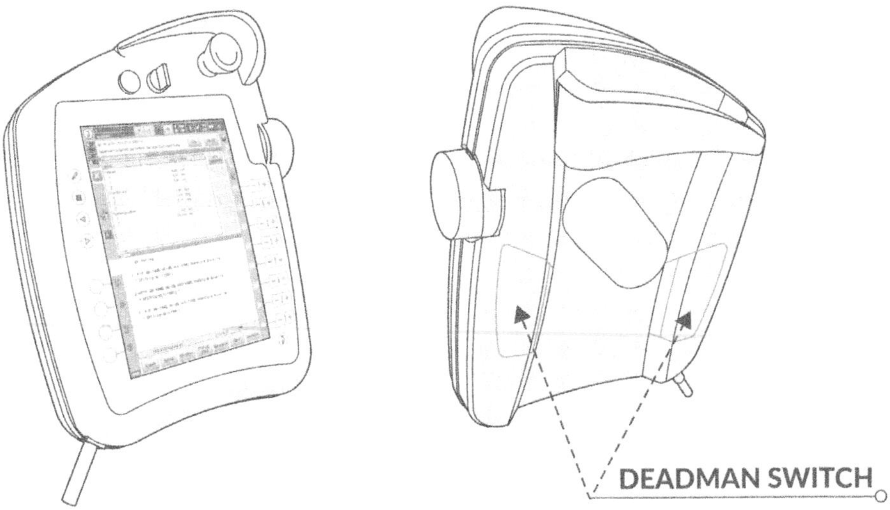

The first control panels resembled huge TV remote controls and were equipped with buttons only. As time went by, more and more displays were added to make programming and diagnostics easier. The latest trend is to reduce physical button to an absolute minimum. Manufacturers try to take care of ergonomics, so that many hours of work with the operator's panel is as comfortable as possible. The perfect control panel should allow comfortable work for right and left-handed people, it should be lightweight and have buttons marked in a way that makes it easy to understand for people of different nationalities. More and more of them are starting to look like tablets, equipped with large touchscreens and user-friendly interfaces. In order to protect them from the effects of a fall, they are often put in special rubber or foam protective covers.

All operator's panels are equipped with a 3-position deadman switch. In the manual mode, it allows the robot to move only when it is in the middle position (partially pressed). If it is fully pressed in or completely released, the robot stops. This principle of operation allows natural reflexes to be used to increase safety. Most people squeeze their hands out of fear when they suddenly notice an incoming robot. The same reaction can be observed when an electric shock occurs. This activates the safety system and in many cases avoids collisions. In the event of unconsciousness or injury, the grip is usually released, which also leads to robot being stopped. Making it necessary to press the button halfway down additionally protects the operator from the robot starting to move unexpectedly.

All safety aspects are primarily regulated by legal provisions, safety instructions from robot manufacturers and factory's health and safety standards. It is the responsibility of the programmer and of the people working with industrial robots to obey those rules at all times. More information on the basic aspects of working with robots in a safe way can be found in chapter 7.

Safety is discussed in a separate chapter, but this is the moment where I would like to include some elementary information about the robot's built-in safety features. Each operator's panel is equipped with a red emergency stop (e-stop) button, which will initiate breaking of the robot immediately after pressing it. As we already know, unfortunately this does not mean that the robot is stopped instantly. The emergency stop button can be deactivated by turning it in the direction of the arrows drawn on it.

Fig. 22: Emergency stop button.

Industrial robots can operate in two modes: test mode (manual) and automatic mode. Operating mode can be selected via the key located in the control cabinet's mode selector switch or on the operator's panel.

Tab. 2 Industrial robot run modes.

Markings	Mode	Maximal speeds
T1	test	manual control (programming): 250 mm/s execution of the program: 250 mm/s
T2		manual control (programming): unavailable program execution: unlimited, speed according to the program
AUTO	automatic	manual control (programming): unavailable program execution: unlimited, speed according to the program

Robot's movement speed in T1 mode is limited to a maximum of 250 mm/s. This limit applies both to manual operation of the robot during programming and to program execution. Robot's program can only be started from the operator's panel and only if the 3-position deadman switch is pressed halfway down.

The second test mode (T2) does not offer the possibility to control the robot manually (programming). The programs can be run at the speed declared in the robot's motion instructions (i.e., in most cases – the maximal speed the robot can move at). It is still necessary to push the three-position deadman switch to the central position. In some countries (USA, Canada) this mode is not available due to legal regulations.

Robot program in automatic mode is executed at the speed declared in the motion instructions – it is used for production in automatic mode (without human supervision). Similar to T2, this setting does not offer the possibility to control the robot manually. It is required to switch to T1 mode for programming. Programs can be executed without the operator, but no one can stay at the station during this time, all people must be outside the fence. In order for the robot to operate in this mode, external safety systems must be active (e.g. closing the door to the station is monitored, making sure that nobody can accidentally enter it). The 3-position deadman switch is inactive in this mode and does not need to be halfway pressed in.

Since 2006, safety standards allowed the use of wireless operator's panels. Until now, Comau and OTC Daihen have decided to use such solution, but the standard for the market still remains the same where the operator's panel communicates with the cabinet via wired connection. There are many additional factors that need to be taken into account when designing a device that uses wireless connection. Communication must be reliable and electromagnetic waves are susceptible to interference and can be suppressed by obstacles in their path. The signal can be easily drowned out by interference from other devices. Security reasons require no delays, which is particularly important for the transmission of safety data (e-stop). Radio signal is much easier to be intercepted and listened to by unauthorised parties. Additionally, lack of a cable means that there is no external power supply, forcing the use of batteries with a capacity allowing for autonomous operation for a long time. This has a direct impact on the choice of the transmission standard, which must be energy efficient, while offering the appropriate range and bandwidth. Pairing of the devices should be convenient and safe. It is also important to bear in mind the availability of free frequencies and the potential need for a radio licence, which may be required in some countries.

When designing their control panel, Comau decided not only to free the user from the cables, but also to take advantage of the additional possibilities offered by wireless technology. Controllers from different robots can be monitored and coordinated using a single operator's panel. It also works the other way round – several operators can supervise one robot control cabinet. The mechanism is based on the Wi-Fi standard (version 802.11 a) due to a number of advantages: range up to 100 meters, transmission security, no need to hold a license for this frequency range. The communication protocol is a proprietary solution patented by Comau [10].

Cables from the operator's panels have the same annoying feature as any other – they like to twist and get entangled. To make life easier for programmers and operators, a solution known from home vacuum cleaners has been adopted. Independent producers and some of the robot manufacturers offer optional cable reels that make it easy to keep cables in check (products of Robot Reels LLC or ABB FPRC – Flex Pendant Retractable Cable System).

3.4. Robots with serial kinematics

General structure of the serial kinematics robot is covered at the beginning of this chapter. This is the most versatile and common type of robot. This is no coincidence – such system can offer large working space with small dimensions of the manipulator itself, including robot's ability to reach behind itself (without the need to rotate on the first axis). This last option sounds particularly interesting, although it is usually limited by the additional equipment mounted on the upper arm of the robot.

Serial robots use rotational joints, which are much easier to seal than linear ones. It is possible to reach a selected point in space in several ways (with different arm configuration), which makes it easier to avoid obstacles and collisions. The disadvantage of that is low rigidity, especially when the robot is stretched out.

A great number of applications in which serial kinematics robots are used has made it impossible to make optimal use of one arm type for all tasks. That is why several variations of these robots have been developed over the years. Nowadays they are adapted to the specific requirements and available in a wide variety of designs. In the past, the robot's reach and lifting capacity went hand in hand, while today manufacturers offer robots with very small ranges compared to the lifting capacity (KUKA Quantec Nano family) or, on the contrary, robots with low lifting capacity and very long arm reach (e.g. Fanuc M-710iC/12L). In addition, there are versions designed for particularly demanding environments.

3.4.1. Shelf-mounted robots

When designing modern and automated production lines, it is important to ensure that robots are placed as densely as possible in the production hall. A large number of manipulators work together to increase production efficiency and flexibility. Distributing work between two or more robots considerably reduces cycle time. Unfortunately, the possibilities of increasing the number of simultaneously operating manipulators are very limited.

Let's analyse the process of welding a car body when a complete bodyshell is already built (the sides of the car, the underbody and the roof are connected with each other). Modern passenger cars range from about 3.5 m to more than 5 m in length. Six robots (three on each side of the vehicle) is the standard quantity that allows optimal use of the installed equipment. Further attempts of adding additional manipulators caused a situation in which some of them, instead of working simultaneously with others, must stand still and wait for the free working space. Instead of increasing the productivity, we have decreased it (and caused additional costs in the form of an idling robot). The solution is to place the robot on a platform that will be placed above the body (a clip showing the assembly of a car body: roboticsbook.com/shelf/).

This requires modified design devices that have a significantly increased working area below their base. Manufacturers have achieved this by reconstructing the base of the standard robot (mainly the second axis) and changing the arm design. This type of robot has a much larger vertical range, as shown in Fig. 23.

Now let's take look at a catalogue and see what a typical spot welding robot looks like. As shown in Fig. 23a, the ABB IRB 6700-245/3.00 model can reach slightly beneath itself (according to the documentation: 596 mm). Analogical (in terms of lifting capacity and arm length) robot intended for platform mounting is marked as IRB 6650S-200/3.0. The changes allowed to achieve a significant increase in the range below the robot's base – 2177 mm (Fig. 23b). This result is by 1.5 m better than the result of a standard robot [8] [11]. When the welding torch is installed, it is practically possible to make welds that are more than 2 m

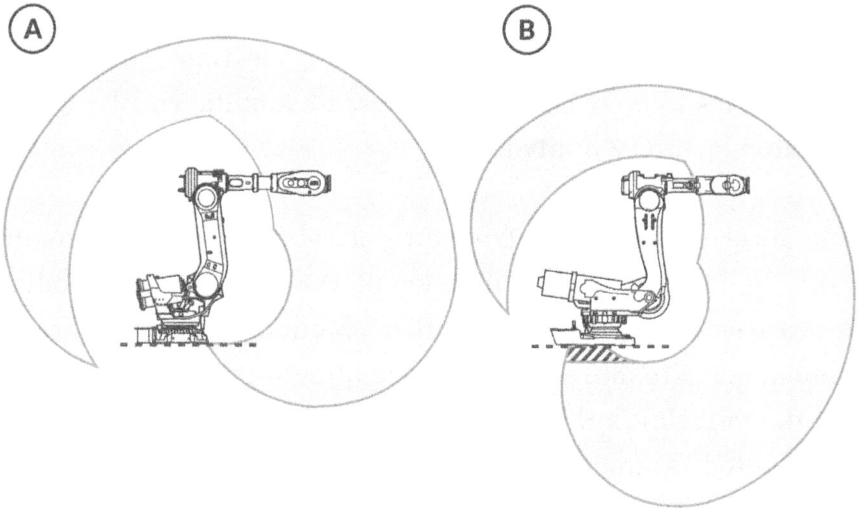

below the platform where the robot is placed.

Fig. 23: Comparison of shelf-mounted robot's and classic robot's working space shape.

Going back to the issue of increasing the density of robots in the hall, it is worth noting that a new category of compact, high-capacity robots has recently become available on the market. They have short arms and a very good ratio of own volume to lifting capacity, which allows them to be mounted very close to the production line (ABB IRB 6620, KUKA KR 120 R1800 nano, Motoman MS120). In this way, we can place robots on the sides of the car's body in two rows. Compact models with low height will be responsible for making welds within the underbody area, while regular manipulators will handle the remaining points on the roof and sides of the car.

3.4.2. Portal robots (Gantry)

Portal robot (Gantry) is a type of robot which has its arm suspended over the ground. These include devices with simple kinematics, resembling overhead cranes, as well as more complex structures using arms with more degrees of freedom. In the latter case, we are dealing with something that resembles a typical six-axis robot, which is mounted on a suspended rails supported by steel poles. Despite its visual similarity, it has five rotary axes and one sliding axis. Robots of this type are usually used to operate machines (lathes, machining centres), and because they are suspended, over the ground level they do not block access to those devices.

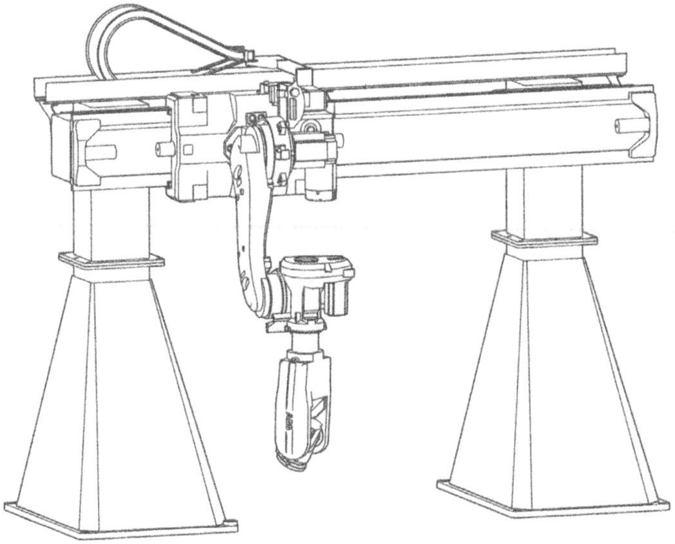

Fig. 24: Gantry robot.

Axis number one (A1), instead of rotating, carries out a linear motion. This combination makes it possible to offer very large working spaces. Compared to conventional robots mounted on the seventh axis, portal robots have more compact design, and in comparison with the shelf-mounted robots they have a lower payload capacity and a larger working space. Balance of forces acting on the rail where the robot runs is much less advantageous as it is a floating structure which must be characterised by the proper precision of movement. It is not possible to carry the same loads as in the case of a free-standing shelf-mounted robot placed on a solid steel structure. Such action would reduce the accuracy and service life of the gear unit and the first axis drive. ABB IRB 6620LX is an example of a Gantry robot, which can be viewed under the following link roboticsbook.com/gantry/.

3.4.3. With hollow wrist

For a robot to perform any task, it must be equipped with the right tools. These, in turn, must be supplied with a variety of media – compressed air, water, electricity, control signals – in order to operate properly.

In standard form, they are delivered to the tool via cables which are routed outside the robot in a protective plastic tube (Fig. 25). It is a simple solution from the structural point of view, but unfortunately it is exposed to many risks during work. There are a lot of components in the robot's environment that can be hit by the cables. There are situations where the robot crushes the cables against itself or they catch on the protruding parts of the tool that is mounted on it.

With the correct selection and installation of cables outside the robot, their durability does not have to cause the maintenance staff lose sleep over them, although they require constant care and attention. Renowned robot cable suppliers offer a wide range of products for different robot models, suited to the specific operating conditions and potential risks.

Fig. 25: Standard way of connecting media to the tool mounted on a robot.

The problems are usually caused by poorly selected cabling parameters. Proper installation has a great influence on their lifespan. The cable must be able to move freely, but it shouldn't be too loose. If the cable is too long, it will catch on the elements of the station while the robot is moving and will be subjected to higher dynamic forces. Similarly, if it is too short and tight, this would also negatively affect its durability. Poorly selected cables are one of the most common causes of production line downtime. The main threats to its durability are continuous bending and twisting of the cable during production, as well as friction and abrasion. Robots working on a production line for a longer period of time can be easily recognised by the fact that the paint on their arms is worn to the bare metal in places, where it comes into contact with the cables. Moreover, harsh operating conditions occur in some applications (welding, metallurgy), where cables are exposed to liquid metal splinters and high temperatures.

Specially designed manipulators with hollow wrist offer the possibility to lead the cables inside their arms. This allows the robot to work in tighter spaces without having to worry about damaging the cables.

They have a longer service life because they are under much less strain than if they had been routed outdoors. The cable placed inside the arm has a limited contact with the external factors, and its construction is selected by the manufacturer so that the cabling life is as long as possible. The price paid for higher maintenance-free operation is a limited motion range on the end axes (usually fifth and sixth), as shown in Tab. 3.

Tab. 3: Comparison of the motion ranges of robots with different wrist designs [12].

Robot axis	ABB IRB 6640	ABB IRB 6640ID (with a hollow wrist)
A1	+170 ° / -170 °	+170 ° / -170 °
A2	+85 ° / -65 °	+85 ° / -65 °
A3	+70 ° / -180 °	+70 ° / -180 °
A4	+300 ° / -300 °	+300 ° / -300 °
A5	+120 ° / -120 °	+100 ° / -100 °
A6	+360 ° / -360 °	+300 ° / -300 °

According to the data gathered by Comau, the average uptime of their robots (without taking into account the cabling life) is equal to 70,000 hours. Meanwhile, the durability of cables routed outdoors (with continuous care and proper installation) is about 30,000 –40,000 hours. As a result, the robot with the external wires and hoses runs smoothly for an average of 20,000 to 28,000 hours. Comau's tests for hollow wrist models, on the other hand, show that the average cabling life inside their robots is equal to 3,375,000 hours. Its' reliability is significantly higher than that of a mechanical unit. In total, the MTBF for a robot with a hollow wrist (including internal cabling) equals to 69,000 hours [13].

Nowadays, hollow wrist designs in practical are a standard solution for robots with low payload capacity intended for arc welding. The implementation of this solution for robots with a payload capacity of over 100 kg posed a bigger problem. Fortunately enough, almost every leading manufacturer has heavy robots with a hollow wrists in their offer now.

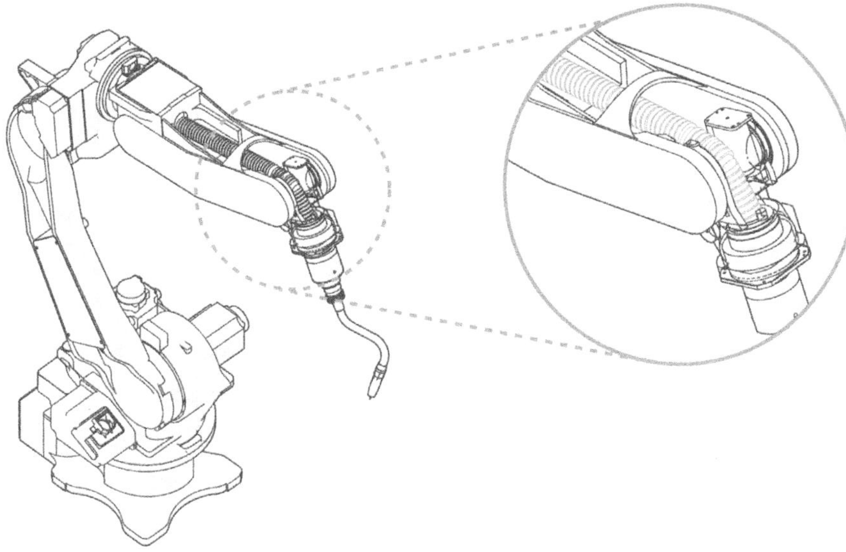

Fig. 26: Robot with a hollow wrist designed for arc welding.

The cabling inside the robot's arm makes the programming process simpler (especially in areas where the robot has a very limited space). Performing a realistic simulation of cable behaviour is a complex task, so it is never known whether an offline-generated program will need to be corrected due to an unplanned cabling placement. The lack of external cables allows the robots to work closer to each other, which makes better use of the space available and allows for an easier programming. The robot's working area is widened and there is no risk of crushing the cabling with the robot's wrist being put in an extreme position.

3.4.4. Cleanroom robots

Today's electronic systems are produced in such low-value technological processes that a speck of dust can ruin the whole system. In order to automate production in such places, it was necessary to develop robots that would not cause additional contamination in a sterile environment. Thanks to design changes and additional seals, it is possible to meet the requirements of ISO3 class 1. As a standard, the number of particles of 0.1 μm must not exceed 10 per cubic metre of air in a room of this class. Cleanroom robots are used in the electronics, pharmaceutical and biotechnology industries.

There is an increase in the popularity of using cleanroom robots in hospitals. The Artis Zeego Angiography machine is a KUKA robot in a special casing and an integrated medical imaging head. Industrial manipulators are also used in oncology wards. Thanks to their precision, they are able to direct a precise beam of radiation exactly into a tumour cell.

Tab. 4: Classification of the quantity of permissible contaminants [14].

Class	particles/cubic metres					
	0.1 μm	0.2 μm	0.3 μm	0.4 μm	1 μm	0.1 μm
ISO 1	10					
ISO 2	100	24	10			
ISO 3	1,000	237	102	35		
ISO 4	10,000	2,370	1,020	352	83	
ISO 5	100,000	23,700	10,200	3,520	832	
ISO 6	1,000,000	237,000	102,000	35,200	8,320	293
ISO 7				352,000	83,200	2,930
ISO 8				3,520,000	832,000	29,300
ISO 9				35,200,000	8,320,000	293,000

Places where food is produced constitute a specific group of rooms with a stricter purity regime. Robots are widely used in food production – from standard applications for sorting food products or palletising food to exotic examples such as frying meat for burgers or cutting half-carcasses. Any contamination coming from the robot must not be harmful to a humans.

When the robot is in direct contact with food, manufacturers offer special versions that use food grade lubricant (NSF H1) to lubricate the drives and gearboxes. According to the American National Sanitation Foundation (NSF) classification, products marked as NSF H1 are food-grade lubricants and their incidental contact with food is acceptable. The maximum permitted concentration is up to 10 mg of lubricant per kilogram of food. Interestingly enough, in addition to being allowed to come into contact with food (lack of heavy metals, carcinogens, etc.), lubricants manufacturers ensure that their products are kosher and halal certified, which means that an accidentally contaminated food does not violate religious orders observed by Jews and Muslims. Even if the leakage of the robot's lubrication system would to occur, no harm would come of it. These manipulators are also resistant to cleaning with chlorine-containing fluids and quaternary ammonium agents, which are commonly used in the food industry for disinfection purposes.

KUKA devices in Arctic version are an example of an interesting series of palletizing robots. The market for frozen products and ice creams in Europe and the United States represents a large part of the food industry. In response to its needs, robots were built to work in freezers. They were designed for use at low temperatures since the very beginning. This involved the use of different types of seals, lubricants, changed tolerances and materials. The media supply cabling has been redesigned to be flexible at low temperatures. Parts can be replaced and repaired by workers wearing gloves. Standard control electronics tolerates temperatures below 0°C very poorly, which is why special solders are used, which do not crack at such low temperatures. A typical KUKA Quantec Prime operates in a temperature range of 10 to 55°C and, thanks to the modifications, its "Arctic" version can operate in a temperature range of -30 to +5°C.

3.4.5. Foundry robots

When discussing sterile clean rooms, it is also worth mentioning the extremely dirty. The ones with high dust content, high humidity and flammable substances. Special robot versions are available for heavy duty applications for installation in foundries, rolling mills, glass and ceramics factories. An increased pressure is maintained inside the robot to prevent dirt from entering the manipulator. The machines operating in highly corrosive and humid environments are covered with additional coatings that increase their resistance to rust. The paints used for painting are resistant to high temperatures, reaching up to 200°C. The robot's wrist is usually silver, which limits its heating. Additionally, it is made of a material more resistant to thermal stress. Robots modified in this way can operate in both acidic and alkaline environments. A robot working in these areas must be able to cope with splashing metal droplets and contact the coolants and oils they use. In order for a robot to work in explosive environments, they must comply with the directive 94/9/EC (ATEX).

Information on the dust and water resistance of the robot is provided in the catalogue notes by means of the IP marking (International Protection). An IP level is usually composed of the letters "IP" and two characters (up to four). The first character specifies the protection against solids and the second character defines the resistance against water penetration. Some robot models have higher levels of protection for the wrist compared to the rest of the arm.

First character (IPx0): protection against solid objects.

Tab. 5: IP Markings – first digit

Level	Protection type
0	No protection
1	protection against bodies with the size of more than 50 mm (accidental hand contact)
2	protection against bodies with the size of more than 12.5 mm (accidental finger contact)
3	protection against bodies with the size of more than 2.5 mm (accidental contact with wire or screwdriver)
4	protection against bodies with the size of more than 1 mm (thin tool, thin cable)
5	protection against the dust penetration in such quantities as to disturb the operation of the device
6	dust-proof – complete protection against dust penetration

Second sign (IP0x): protection against water penetration.

Tab. 6: IP Markings – second digit

Level	Protection type
0	No protection
1	protection against vertically falling water drops (from condensation)
2	protection against water drops falling on the casing at an angle of 15° to the normal position
3	protection against water drops falling at an angle of 60° from the vertical
4	protection against water drops falling at any angle from all sides (rain)
5	protection against water stream from any direction
6	protection against strong water streams or flooding from any direction
7	flooding protection during immersion to such depth that the lower surface of the casing is 1 m below the water surface and the upper surface is not less than 0.15 m for 30 minutes
8	protection against flooding by continuous immersion and increased water pressure (1 m deep)
9	protection against flooding by a pressurised water stream (80-100 bar, temperature up to +80 °C)

The market offers industrial robots with a very high level of protection, up to IP67 for the whole machine (Denso VS, KUKA KR Agilus WP, Fanuc R-2000iC/210WE), which, theoretically, could operate while completely immersed under water without any damage.

The quality of seals used in industrial robots can be proven by the fact that one of the applications of ABB IRB 6650S robots is washing of haul trucks working in open pit mines at Carajás mine in northern Brazil, where iron ore is mined. The environment in surface mines is dirty, dusty and corrosive, which causes rapid degradation of dump trucks. Manual cleaning takes four to five hours, and can be carried out only during the day. It is a strenuous task, but robotisation removes all these inconveniences. Robots have been working in Brazil since 2007 (for more information see: robotiсsbook.com/cleaning/).

I have never encountered any practical implementation of any process where an industrial robot would dive under the surface of the water. Watches are similar – only those with water-tightness of 50 meters or more allow swimming (without diving) in an ordinary swimming pool. It is caused by the fact that during movement the dynamic pressure of the water acting on the seals exceeds the permissible values (which are measured in static conditions). Nevertheless, the manufacturers at the trade fairs are keen to boast about the capabilities of their machines and one can see there a robot arm emerging majestically from the surface of the water and continuing to perform a demonstration trajectory.

3.4.6. Palletising robots

Palletisation is a simple process that is easy to automate. Due to the high monotony of work and, oftentimes, the need to move heavy loads, companies are keen to use robots, which speed up the handling process and save space at the factory. People's working conditions are improving and the number of accidents are on the decline. Manufacturing plants often start with palletising in order to robotise production. These positions are characterised by a high rotation of staff and, usually, by accidents and medical conditions caused by long-term overworking one group of muscles and joints. Robotisation of such tasks does not require a lot of money (the stations are not very complex – robot, gripper, feeder).

Usually robots with four axes are used for palletising (although in extreme cases this number can fall to as much as two: e.g. Fanuc M-421iA). This is a sufficient number of drives to move the components from one place to another. By reducing the number of moving parts, the arm stiffness and the load can be increased with the same dimensions. In the case of designs with a reduced number of axes, it is important to ensure that the area where the robot is going to be installed is precisely levelled. It will not be possible to correct potential deviations afterwards – the mounting plate of the four-axis robot moves perpendicularly to the surface of which it is mounted on.

Parallelogram mechanism is commonly used in palletising robots. It is in the form of an additional rod which carries a part of the load and stiffens the arms at the expense of freedom of movement (it is not possible for the robot to reach "behind", this area is available only after the rotation of the whole unit on the first axis). For all other tasks this would be disadvantages, but for palletising this type of mobility restriction is not important. The use of a parallel mechanism relieves the robot's third axis and its drive can be placed in its base, which results in a slimmer construction of the arms. The comparison of the robots with and without a parallel mechanism (ABB IRB 660-180/3.15) is shown in Fig. 27 – the first robot has a significantly smaller working space.

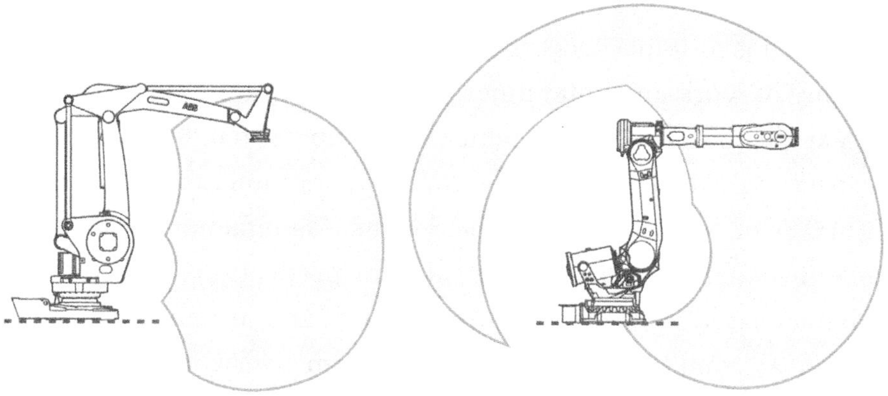

Fig. 27: Influence of the parallelogram on the working space of the manipulator.

The number of cycles per minute constitutes an additional parameter for the comparison of palletising robots. The measurement is usually performed by moving the robot along the trajectory shown in Fig. 28. However, the large scattering of the load capacities and the robot ranges means that each manufacturer determines their own lengths and loads. It is difficult to talk about a common standard.

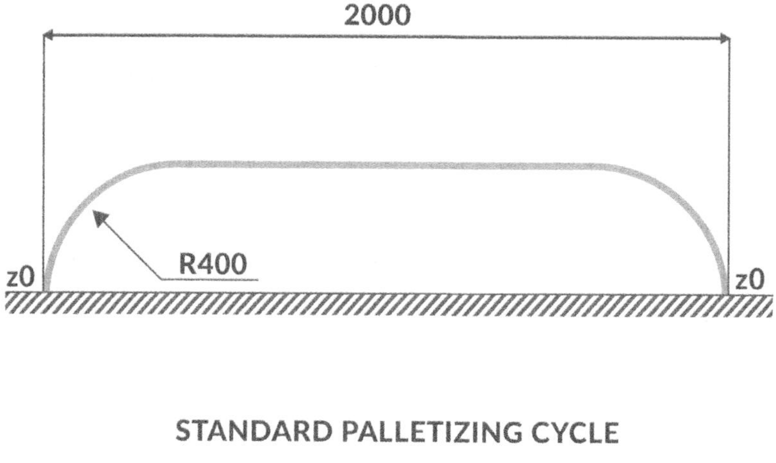

Fig. 28: A trajectory as a benchmark for palletising robots.

A very common option for robots in this category is to equip them with a mechanism to synchronise the robot's movements with the conveyor tracking. This technique allows products to be picked up while the conveyor is in motion, without the need halt production. The robot is capable of calculating the distance which the products will cover during the picking up and setting down of the item based on the information inputted by the operator.

3.5. Parallel robots

All the robots discussed so far have a serial structure. Parallel robots appeared on the market quite late, as it was not until the 1980's that they were introduced. Parallel manipulators are mechanisms with a closed kinematic chain, of which the base is connected to the working platform through independent arms. Manipulators with hybrid structure, offering greater versatility, are also used in the industrial practice. The advantage of parallel robots is the level of high accuracy they provide, which results from the fact that positioning errors on individual axes do not add up with each other, as opposed to serial mechanisms. The only disadvantage of this solution is its complicated mechanical construction, which requires an advanced control system.

Delta robots represent the most recognisable devices of that type. The idea of this robot was born in the 1980's in Switzerland. Professor Reymond Clavel of the École Polytechnique Fédérale De Lausanne (Lausanne Federal University of Technology) proposed to use a three-parallelogram structure to build a robot with four degrees of freedom (three shifts and one rotation). He protected his idea with 36 patents. The most important of which was *Device for the movement and positioning of an element in space*, which was granted in 1987. The first devices were produced by the Demaurex brothers Marc-Olivier and Pascal, who bought licences for the production of Delta robots in the same year. Currently, these patents have expired (in Europe in December 2006 and a year later in the United States), which resulted in the emergence of many solutions based on this idea, but built by various manufacturers.

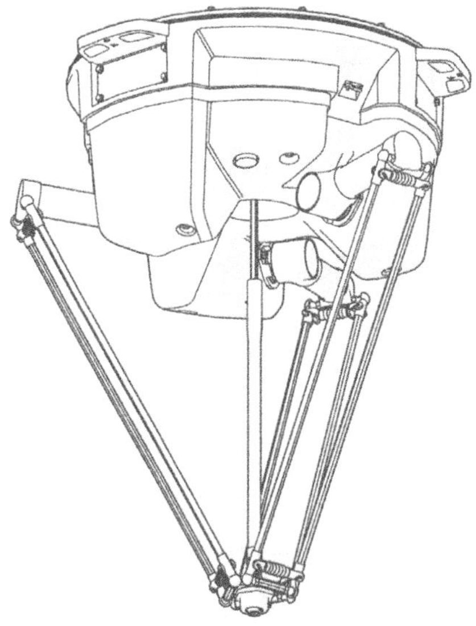

Fig. 29: Parallel robot manufactured by ABB.

Delta robot is a closed-loop kinematic chain type robot. The parallelogram elements are connected by ball mechanisms with three degrees of freedom. Delta robots are used quite frequently in the electronics, chemical and food industries, where their main task is to handle and sort lightweight parts. They are characterised by low payload capacity and a very high movement speed. This can be achieved by mounting the drives in the robot base and using lightweight arms (with low moment of inertia and high rigidity). Their disadvantage is the small working space. As with conventional robots, it can be increased by mounting the robot on a linear motion track.

Tricept is another type of a parallel robot (in principle, they are considered hybrid kinematics devices – part of the robot is parallel and part is serial). Patent concerning such machine with parallel kinematics (PKM) was awarded to Karl-Erik Neumann in 1985. The cooperation between Comau and Neos companies in 1992 resulted in the development of the first product on the market – Tricept TR600. Neos has evolved into Tricept AB over time and has been known as PKMtricept SL since 2004. They supplied arms to Comau who used them and offered their own products (under the name of HP1) and ABB (model IRB 940).

Tab. 7: Comparison of parallel and articulated robots' characteristics.

	Articulated robot	**Parallel robot**
Workspace	large	small
Maximum force	limited by the weakest drive	sum of the forces of all drives
Rigidity	low	high
Inertia	high	low
Payload/robot's weight ratio	small	high
Workspace/robot's size ratio	high	small

Parallel robots with high payload capacity are becoming increasingly irrelevant. They are too slow and at the same time they have too little working space to compete with the next generation of articulated robots. This does not mean, however, that you cannot find interesting applications for them. Ford has developed Ford Freeform Fabrication Technology, a device that allows metal sheets to be formed flexibly without the need for embossing. This facilitates rapid prototyping and the production of unit test parts (for more information see: roboticsbook.com/ford/).

High rigidity and high payload capacity with good accuracy constitute one of tricept robot's advantages. The disadvantage, as with the Delta robots, was the limited working space and the unfavourable ratio of the robot size to its working area. These manipulators have been used for heavy-duty, high-pressure and high-precision applications, such as friction stir welding (FSW).

3.6. SCARA robots

The SCARA robot was invented by Hiroshi Makino and its design was inspired by Japanese folding screens. This type of manipulator owes its success to the change in approach to the method of manufacturing devices. Before the SCARA robots appeared, the designers created products that were difficult and complicated to assemble. Their manufacturing process involved a relatively small number of complex operations, which required advanced robots with many degrees of freedom.

The Japanese proposed a different approach – the assembly of the devices should be broken down into many simple steps. This requires more production machinery, but thanks to their simple structure, costs less. There is an additional advantage to this approach – the increase in production flexibility; a high number of robots provides greater freedom in the allocation of tasks. SCARA (Selective Compliant Assembly Robot Arm) was the result of this change in perceiving the assembly of products process.

These robots, thanks to their characteristic arm design, are, to some extent, subject to forces acting on them in X and Y planes, offering high stiffness in Z axis at the same time. This is a great advantage for a robot used in assembly. When the parts that are going to be mounted are not perfectly aligned with each other, they can be assembled together, because the robot can correct the positional inaccuracies by slightly bending its arms. SCARA robots are widely used in the electronics, pharmaceutical and chemical industries. With their simple design, they achieve very high movement speeds and high positioning accuracy.

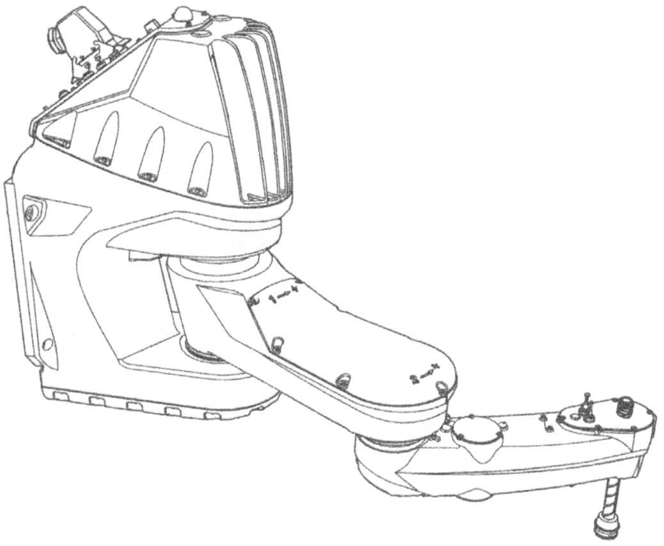

Fig. 30: Stäubli TP80.

Nowadays, the robot's partial compliance to one of the motion axes can also be achieved with conventional articulated robots (functions: Fanuc Softfloat, ABB SoftMove, Motoman Servo-Float). If this option is active, the robot arm will be subject to external pressure by bending over the axes accordingly. You can specify to what level of external force the robot should attempt to execute the program without being affected by that force. Once the pressure stops, the robot may try to return to the programmed trajectory or stay stationary.

Some manufacturers offer classic series structures with five and six axes in place of SCARA robots. Although they may not yet compete with SCARA in terms of speed, their performance is already high enough to take away part of the market, especially where greater arm versatility may be required in the future. SCARA robots compete primarily with Delta robots in pick-and-place tasks. Looking at both types of construction, several common features can be found, such as low payload capacity and very high speed of movement, which makes them equally suitable for use in the same applications.

Tab. 8: Comparison of parameters of SCARA and Delta robots [15][16].

	Stäubli TP80 (SCARA)	**ABB IRB 360-1/1600 (Delta)**
Payload capacity	1 kg	1 kg
Range (horizontal)	1600 mm	1600 mm
Repeatability	±0.05 mm	±0.1 mm
Cycle time 0.1 kg	0.30 s (200 cycles/min) (Path: 25/**300**/25)	0.35 s (171 cycles/min) (Path: 25/**305**/25)
Cycle time 1 kg	0.35 s (170 cycles/min) (Path: 25/**300**/25)	0.40 s (150 cycles/min) (Path: 25/**305**/25)

Tab. 8 contains a list of robots that compete in a common weight category, although they have completely different designs. Stäubli TP80 is a SCARA robot with a currently record-breaking speed in its class (for more information see: roboticsbook.com/scara/) with low payload capacity (max. 1 kg, nominal 0.5 kg) and large working area. ABB IRB 360-1/1600 represents Delta robots, and it is slightly slower than the first manipulator in the table. Compared to the competition, it offers relatively low repeatability, but it is also available in versions with higher payload capacity.

3.7. Collaborative robots

For many years, Asia has been the place where European and American companies have relocated their factories to benefit from cheap labour. If costs had to be reduced, the first idea was to relocate production to a cheaper part of the world. The global economic situation has changed significantly in recent years. The salaries in China are growing more dynamically than in the Old Continent or the United States. The one-child policy conducted between 1987 and 2003 reduced the number of new employees entering the labour market. The amount of support given by the Chinese government to foreign investors is also decreasing.

At the same time, the product quality requirements are becoming more and more strict. In many cases manual production proves to be too inefficient and not precise enough. Companies had to invest in automation to meet customer demands and attract new contractors. An increase in the widespread of automatic operations results in independence from the labour costs of workers. It turns out more and more frequently that despite higher costs of human labour, a greater profit comes from the production on the domestic market.

The cost of buying a robot is more or less the same in every corner of the world (and the possible differences are due to the amount of duty and tax). It is also a "one-time" expense – once the machine has been bought, it works for many years, which means that the costs of its purchase in the context of its entire life cycle are not so significant. What is more important is the availability of specialists capable of servicing them, the presence of service facilities and manufacturer's support. Western countries offer easier access to experts who can install, program and service robots. When setting up a factory in Asia or Africa, one must take into account the necessity of sending workers there (who are paid either in euros or dollars and not in yuans or rupees) to commission lines and train local staff.

Sending specialists and paying for their allowances and hotels is a considerable financial burden. A plant located in Asia, Africa or Central and Eastern Europe, in a situation where most of the customers are located for example in the United States, constitutes a source of additional costs related to logistics. Transportation time for manufactured goods takes weeks, not days or hours. This makes it impossible to manage the demand for products on an ongoing basis. It is also related to high delivery costs and the risk of damage to the cargo on its way. Many companies are beginning to realise that production in China and sales in their home countries are becoming less and less profitable. Therefore, in the upcoming years, an increasing number of American and European companies will invest in factories located in their home market. Caterpillar, Boeing or Ford are already moving part of their production back to their home market – the United States. Economists and robotics distributors in this country are using the term re-shoring (moving factories from Asia back to the United States) in all its forms.

Labour costs are not the only thing related to production, it is also the energy prices, stability of regulations and taxes. Customers are increasingly inclined to choose their local made products, even if this entails a slightly higher price. They know that by subsidising domestically produced goods, they are driving the economy and maintaining jobs.

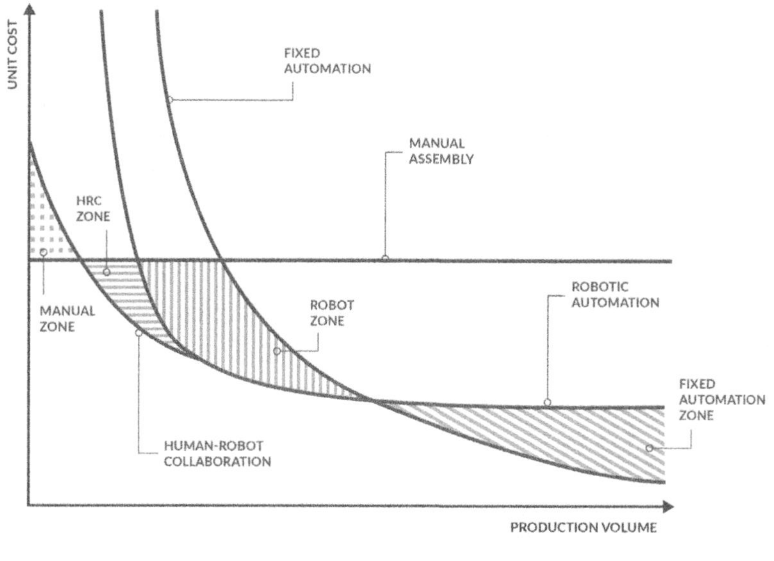

Fig. 31: Field of application for HRC collaborative robots.

These problems are a nightmare for large corporations, but also for small businesses who are also facing the problems of an increasing competitiveness. These companies, unlike mass producers, to a much greater extent base their production on the work of human hands. Thus, their profitability is much more dependent on labour costs, and the temptation to move production to Asia, Africa or Eastern Europe is much stronger. Popular clothing brands, which outsource their production to their Asian and African partners, are a perfect example of that. This has a negative impact on the economies of developed countries – skilled workers are losing their jobs, the unemployment rate is rising and GDP is decreasing.

Perhaps one should follow the example of the American and European companies – to robotise production and thus become largely independent of human labour costs? Until recently, it was a purely a theoretical solution. Only a few years ago, robot manufacturers had nothing interesting to offer small and med-size enterprises in their catalogues.

Small companies manufacture products in small series and a large variety. It is economically unreasonable to use an expensive industrial robot that requires a long-term configuration, and employment of a specialist programmer or hiring off his services. The robots are designed to be fast, efficient and trouble free. Matters such as easy installation or quick programming were of secondary importance a few years ago.

Moderate velocities and forces, inconsistent and poor quality output – these are the limitations of human labour. Robot is better in all of those cases. However, it gives way to man in terms of adaptability to change, autonomy, mobility or the size of the working space.

Tab. 9: Comparison of features of robots and people.

Advantages of robots	Limitations of human workers
▸ high speed ▸ high strength ▸ high repeatability ▸ constant quality	▸ moderate speed ▸ moderate strength ▸ poor repeatability ▸ high quality dispersion
Robot limitations	**Advantages of human workers**
▸ lack of cognitive ability ▸ lack of self-adaptation ▸ moderate working space	▸ cognitive skills ▸ reacting to change ▸ adaptation ▸ improvisation

The use of robots in small enterprises helps improve their performance on many levels. It is not only more efficient and reproducible, but stands as a significant relief from the employees' workload. Replacing workers in physical activities can significantly reduce the problems of musculoskeletal pains, which represent a common disease in people performing repetitive physical tasks.

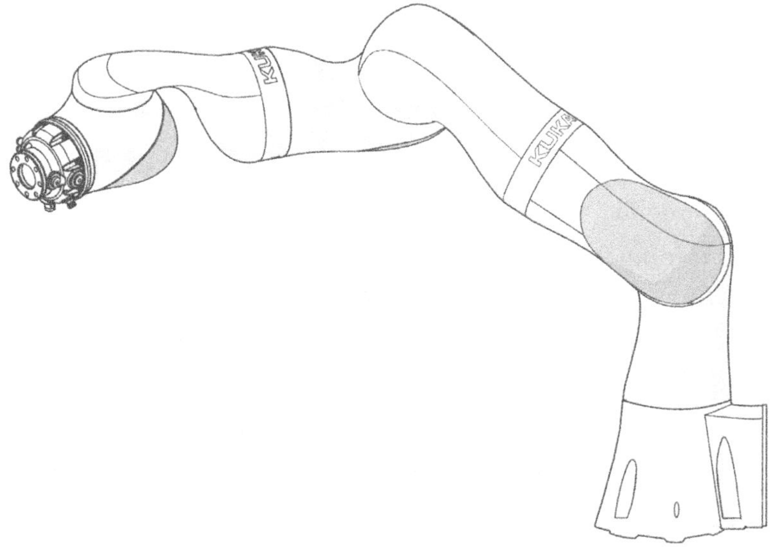

Fig. 32: Cooperative robot – KUKA LBR IIWA.

The European Union, together with robot manufacturers, has noticed the problem of production automation in small companies. A joint SME Robot initiative was established in 2005 (and later SME Robotics in 2012) to develop a family of robots that could support the production process in small companies. It has been agreed that such devices should be:

- safe
- easy to install and program
- flexible (easy and quick change of production profile)
- capable of connecting to consumer computers and equipment
- affordable.

Thanks to the SME Robot and SME Robotics initiative, supported by the European Union, and leading robot manufacturers, it was possible to find solutions that allowed to meet the requirements from the above list. They gave rise to a new family of industrial robots – *collaborative robots* – which are more commonly known as cobots.

The ability to work hand in hand with a mechanical helper required a revolution in safety legislation (ISO 10218, 2006, updated in 2011). Robots should take over small jobs that have been carried out by people in the past – i.e. moving parts, surface treating, drilling holes, assembling or serving as a worker's "third hand". These tasks do not require high lifting capacities, so you can reduce the size of the robot, which has a positive impact on safety. The small size also results in lower energy consumption – in most cases, a conventional 110 or 230 V power supply is sufficient.

Until recently, no one was surprised when the robot looked like a tank – angular, raw arms have never surprised anyone. Collaborative robots are round in shape and are made of soft materials to minimise the possible injuries in the event of a collision. This collaboration requires humans to be in continuous contact with the machine, and rubberised finishes or soft plastics are much more pleasant to touch than a cold, rough casting. In addition, they dampen vibrations, which makes their operation less cumbersome in case of programming by pushing the arm to a given place.

Compared to conventional robots, seven axes are much more common among collaborative robots for greater versatility. Thanks to the additional degree of freedom, the robot can move away from the worker without changing the orientation and position of the held object. It also makes them similar to the human arm, which has the same number of degrees of freedom (three degrees of freedom for each shoulder and wrist, and one degree of freedom for the elbow joint).

Conventional industrial robots have to work in closed areas with no access for people due to safety reasons. If something goes wrong, the robot will stop on the safety fence. Collaborative robots do not have to be as efficient as their larger brothers, but above all they have to be safe. The existence of collaborative robots was sanctioned by the ISO 10218-1:2006 standard, which defined the following restrictions:

- TCP velocity ≤ 0.25 m/s (10 in/sec)
- maximum dynamic power ≤ 80W or maximum static force ≤ 150N.

The ability of limiting the power and force is what distinguishes collaborative robots from conventional robots.

In 2011, ISO 10218-1 was updated and does not currently provide a power/static force limit value. Higher speeds are also permitted. These parameters are determined by the robot integrator, which should select them in such a way to ensure that they are safe. The ISO/TS 15066:2016 standard is under development to fill this gap and clearly define what forces can be used in which cases. It allows you to increase certain parameters of the robot in justified cases (for more information, see chapter 7.1).

Safety functions are provided in a variety of ways. KUKA has equipped its LBR IIWA with torque sensors in each drive to detect increased resistance (collision). Fanuc, in the case of the CR35-iA and CR7iA robots, has redesigned the existing models in its range by adding a soft cover, while collision detection is carried out via a torque sensor mounted in the robot's base. In the case of the Bosch APAS robot, the cover has built-in sensors to detect arm contact with other elements. Robot protection casings are also available with capacitive sensors that detect an obstacle at a distance of a few centimetres – which can be used to avoid a collision.

The works on active safety systems will be able to prevent a hazardous situation from occurring and not only minimise its impact, which is still under development. There are solutions (e.g. safety laser scanners) that can detect people in the work area and stop or slow down the manipulator. In the future, robots should be able to modify their movement path in such a way as to avoid collision.

Although robot controllers have been around for approximately 60 years on the market, they still do not offer the function that is referred to as 'Plug & Play' in the computer world – which means the ability to work with peripheral devices as soon as they are connected, without the need for a user to configure the hardware.

A home computer, in theory, is a device that should not be assembled by an amateur user, because it has a lot of system buses, standards of communication between components (in different versions!) and a variety of connectors. In practice, it is sufficient to connect everything with each other by means of plugs whose unique mechanical shape prevents incorrect assembly. When you buy a new printer and connect it to your PC, your computer should immediately recognise which device it is dealing with. The operating system will communicate with the device and select the right drivers to install for the particular equipment. A device that has just been bought, should start working a few moments after taking it out of the box.

In the case of robots, each newly connected device must be manually and laboriously configured, and the number of communication interfaces does not make it much easier. The number of collaborating robots will increase over time. They will be used in artistic installations, service points and small plants. In the future, it will be necessary to develop interfaces which will make it possible to add a new tool, remove a unnecessary component or perform a hardware update for people who do not have comprehensive knowledge of industrial networks. This will make it easier to adapt the robot to new tasks.

Innovations are supposed to reach not only the way the robot communicates with the periphery, but also the methods of control itself. Research is being carried out into more natural methods of instructing a robot than a teach pendant, including gestures (using the Nintendo Wii console controller), a gamepad from game consoles, or voice commands. Commercial, portable solutions (such as Emotiv EPOC+) already exist for recording brain wave activity and detecting facial expressions. Thanks to the bioelectric recording of brain function, the ability to control a robot with thought does not seem to be a pure science fiction anymore. The programming process itself should also be considerably simplified. There is a returning idea of programming by hand-guiding (see chapter 5.1), allowing you to quickly and accurately plan the trajectory.

The number of robot models that can work in close proximity to a human being is constantly increasing and amongst their representatives one can find Universal Robots UR5, KUKA LBR IIWA, ABB Yumi, Yaskawa HC10. They are easy and safe to use and can work with humans. The cooperation between humans and robots is expected to be one of the most important trends in industrial robotics over the next few years.

Tab. 10: Comparison of parameters of a collaborative robot (KUKA LBR IIWA) with a classic robot (KUKA KR6 R900-2) with similar payload capacities.

Parameter	LBR IIWA 7 R800	KR 6 R900-2
Weight	23.9 kg	54 kg
Payload capacity	7 kg	6 kg
Range	800 mm	900 mm
Repeatability	±0.1 mm	±0.01 mm
Axis number	7	6

Cooperation between humans and robots is not only a technical challenge (development of new materials, drives and control systems) but also a social and psychological one. Classic industrial robots must be fast and efficient. They achieve this goal thanks to very high movement speeds and the possibility of very rapid braking and acceleration. Because they are separated from the workers by a safety fence, they do not give cause for concern to people working nearby. The opposite is true when nothing separates us from the robot, which is the case with collaborative devices.

Based on the facial expressions, tone of voice and reaction of the co-worker, people can predict their future behaviour. We do it subconsciously all the time. If you are trying to analyse robot arms in order to predict its future movements, it may come as a surprise that the only thing you can be sure of are the brand and model markings.

This can be a source of anxiety and subconscious stress. The human body is in a state of constant, increased readiness for unexpected robot movements. Therefore, the control algorithms have to change, the movement has to be smooth, and sudden acceleration must be eliminated. Work on "humanising" the behaviour of robots is still in progress. The conductivity of the skin (depending on the amount of sweat), heart function, pupil enlargement, the electrical activity of the muscles – all of those aspects are measured, monitored and examined by testing different trajectories of movement. In the future, the results of this research aim to create such planners of the robot movement, which will try to execute the manipulator movement in a way that is the least onerous and stressful for people working nearby.

Even today, companies with a high level of production robotisation are trying to warm up the image of robots. The biggest manipulators in the Tesla factory in Freemont are named after the superheroes from "X-men". You can find the Colossus and Storm robots (not the impersonal ir3580r01 and ir3580r02) at the end of the chassis line. Some of the Dragon Ball Z characters have settled in the Nissan factory, for example Son Goku and Vegeta.

3.8. Parameters of industrial robots

Increasing customer demand is a driving force for changes in the manufacturing process. In order to improve the quality and efficiency of production, it is necessary to constantly raise the requirements for the machine park, which also applies to industrial robots. Therefore, it is worth knowing the basic parameters that are important when choosing a robot. What makes one robot better than another?

Each unit has a maximum payload capacity specified by the manufacturer, which is the basic criterion for classifying the manipulators. Payload capacity is the maximum weight that a robot can carry and at which it maintains the parameters set by the manufacturer (positioning accuracy, lifetime of drives and gearboxes). Payload capacity includes the weight of the tool (in the case of a gripper, including the load-held part) mounted on the robot and any additional arm-mounted equipment.

Even if, according to the manufacturer's catalogue, the robot has a much higher payload capacity than the mounted tool weight, its capacity may not be sufficient. The gripper, which weighs 100 kg together with the part, can be too heavy for a robot with a catalogue payload capacity of 200 kg. It all depends on the location of the centre of gravity. As the centre of gravity of the robot attachment moves away from the tool flange, the payload capacity decreases. It works exactly the same as the human arm does. Keeping a 10 kg dumbbell along your torso is not a problem. Problems start when you want to do the same with your arm stretched outwards.

Exceeding the manufacturer's recommended payload capacity will shorten the lifespan of the drives and may damage them. To avoid overloading, you can design a more compact tool (which will bring the centre of gravity closer to the robot), use lighter materials, or use a manipulator with higher payload capacity. In order to reduce the weight of the robot-

mounted devices, welding guns with aluminium arms instead of copper or carbon grippers instead of aluminium are being used more often.
There are two types of overload: static and dynamic. Static overload occurs when the robot's mounted weight generates forces that exceed the servomotor's torque when the robot does not move and cannot hold the robot in position. In the event of dynamic overload, the cargo's inertia exceeds the permissible limits.

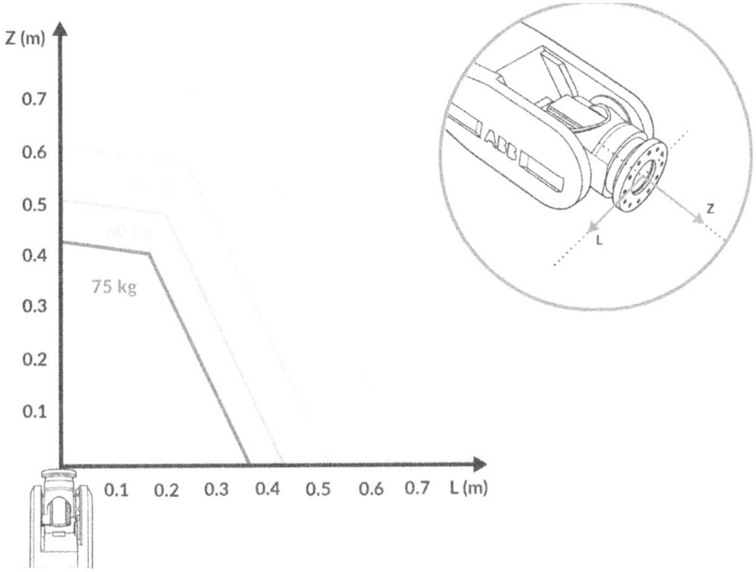

Fig. 33: Permissible load diagram.

The diagrams provided in the technical documentation (such as the one shown on Fig. 33) provide only an approximate indication of the robot's suitability for the task chosen. When selecting a robot, it is strongly recommended to use programs from the manufacturers (*KUKA Load, ABB Robot Load*) to determine whether the load limits will be exceeded based on the information on weight, centre of gravity and moments of inertia of the equipment.

In addition to selecting the correct robot for the assumed working load, it is also important to enter the data of the weight of tools and parts carried by the manipulator into the controller. Knowing the weight suspended on the tool flange, the robot controller is able to control the servomotors in a way allowing to compensate for the inertia forces that occur during the motion. This provides higher precision of movement and increases the life span of the device. Lack of this data or incorrect values may shorten the life of the drives and transmissions, and may cause the robot to stop during the execution of the program due to overloads. With the load information, the controller can plan the trajectory using a dynamic manipulator model to optimise the motion for the best possible cycle time.

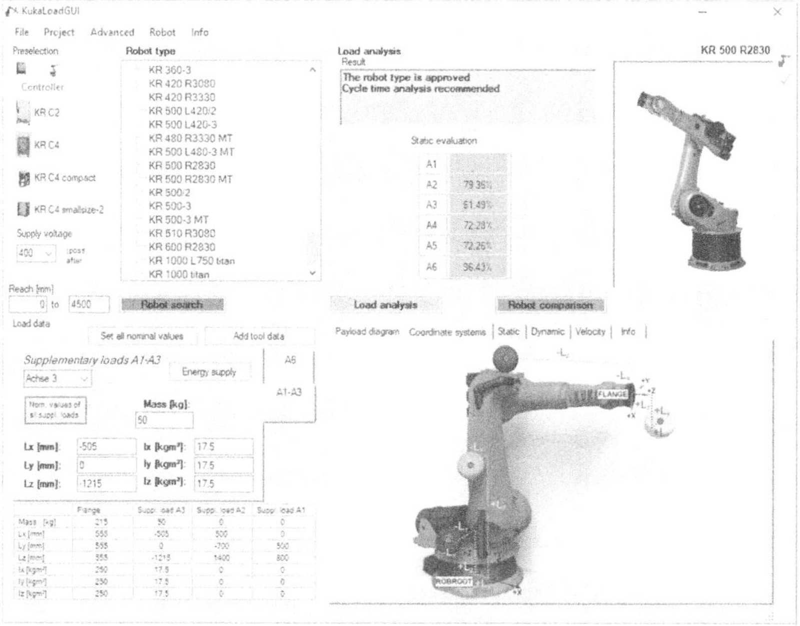

Fig. 34: Window of the KUKA Load program.

If the created program lacked information about the mass of tools installed in the robot, then after adding them, one should take into account that the trajectory of the robot's motion can change insignificantly. Before starting production, it is required to test all programs in manual mode in order to ensure that there is no collision while the line is running in automatic mode. In case of some controllers (e.g. Yaskawa FS100), the lack of information about the load and its centre of gravity will cause the controller to adopt the worst possible scenario and limit the maximum speed in order to protect the durability of the mechanical unit.

Determining the correct load data is extremely important for the trajectory planner algorithm on the controller. What should one do if this information is missing? Robots can be equipped with appropriate software options, which allow the controller to determine these values (mass, centre of gravity, moments of inertia) on its own. The robot performs several measuring movements, during which it analyses the forces acting on the servomotors, on the basis of which the load parameters are determined.

Range is another important parameter – it is usually determined in the vertical and horizontal plane. Both the maximum range (with the arms outstretched to the extreme) and the minimum range (with the robot in the maximally folded position) are provided. For more information on the robot's motion capabilities, see illustrations showing the manipulator's working space. These two parameters, payload capacity and range, are the first and most important criteria for choosing a robot.

When choosing a robot, it is also important to pay attention to the motion ranges of the individual axes and the speeds they achieve. The higher the angular velocity reached, the shorter the cycle times that can be achieved. Typically, the higher the robot's payload capacity, the smaller the robot's range of motion and speed will become. Unfortunately, the analysis of the values in the tables alone is not able to tell us how quickly the robot will cope with the assigned task. For this purpose, we must use simulation software, in which we will create test programs. For processes that require high accuracy (e.g. laser welding), the robot's precision is an important factor. The quality parameters describing the robot's capabilities include resolution, repeatability and accuracy.

Resolution is the smallest shift the robot can make. It is the shortest distance by which the TCP of the robot can be moved.

From the quality of the production process point of view, there are two important parameters of a robot – *repeatability* (RP) and *accuracy* (AP).

Their measurement methods are defined in ISO 9283:1998 (*Manipulating industrial robots – Performance criteria and related test methods*).

Repeatability makes it possible to assess the precision the robot achieves in successive motions to the same location with the same orientation. A robot with a high repeatability parameter will return precisely to its learned position each time.

Accuracy is the ability to accurately reach a point with defined coordinates in three-dimensional space. The difference between accuracy and repeatability is shown in Fig. 35. As a rule, robots can boast much better repeatability than accuracy.

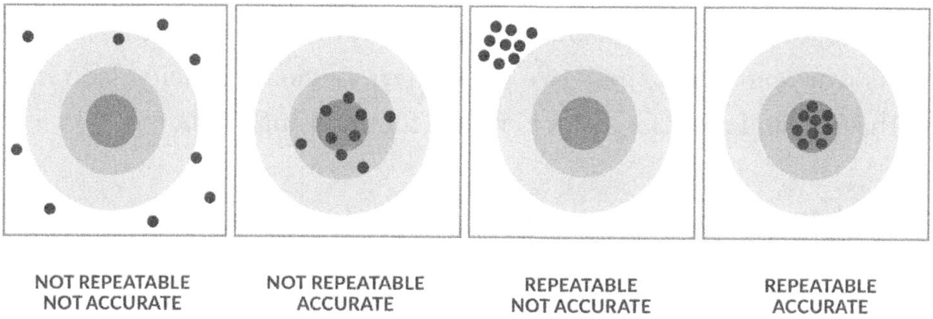

Fig. 35: The difference between accuracy and repeatability.

This is due to the fact that the two industrial robots leaving the production line one after the other are not identical. The differences start at the stage of mechanical processing of the robot arms (which are made with a certain tolerance), the drives and transmissions have backlash as well. In addition, there are nuances such as shoulder deflection caused by suspended loads. This makes the robots of the same type slightly different. In addition, there are details such as the influence of the heat generated by the surroundings and servomotors on the robot's geometric dimensions. Robot arms are usually made of cast iron and aluminium with a thermal expansion coefficient of 10.5 µm/°C and 23 µm/°C respectively, which means that their dimensions change slightly as the temperature rises. In the case of a serial robot, all these imperfections add up, which further deteriorates the accuracy. The robot's controller is unaware of these tiny deviations and has the ideal mathematical model stored in its memory. It does not compensate in any way for deviations from the ideal state, because there is no data about them. Therefore, there is a difference between the actual position of the robot in the 3D space and in the position calculated by the controller from the ideal mathematical model of the robot.

Robots are usually taught by being moved to a point and memorising the position. It is important for the programmer to make sure that the robot reaches the same place in every work cycle. The position value read on the control panel being perfectly in line with reality is usually less important. Operations such as milling and surface treatment are usually treated as exceptions. In case of these, the programs are generated offline (chapter 5.1) and are expected to reflect the reality. If, according to the generated program, the robot is supposed to drive one meter and it covers, in fact, 999.5 millimetres, it may turn out that the detail it has cut will not fit in with the rest of the manufactured device.

A factory calibrated robot (*Absolute Accuracy* option) can be ordered for tasks where very high accuracy is required. Calibration is executed by performing hundreds of measurements of the robot's position, with its various arm configurations and different loads. The actual position results obtained from the external measuring equipment shall be then compared with the theoretical values calculated by the controller and the deviations determined. These measurements are different for each manipulator and allow for taking into account the factors such as mechanical tolerances and arm deflection caused by higher loads. This provides information on robot's imprecision that can be used to improve the positioning accuracy. With the additional information on the recorded deviations, the robot controller has sufficient data to compensate for positional errors. It should be remembered that the versatility of a serial robot is subject to certain limitations, and despite the possibility of improving its accuracy, it will never be able to compete with a CNC machine. This is due to, among other things, lower rigidity and being more susceptible to vibration.

The principle of the *Absolute Accuracy* option is limited to the positions defined in the Cartesian coordinate system. When the robot's position is determined by the angles of the individual servomotors' rotation, it is not possible to compensate for this position, which is caused by the principle of this option's operation. It is a fully software-based solution and after calibration by the manufacturer there is no need for external measuring devices. With its help, the positioning accuracy in the whole manipulator space falls within the ±1 mm range.

Robotics have not escaped from ecology and green technology. In recent years, user-cost issues have become increasingly important when selecting a new robot. These include the cost of spare parts, the ease and the time it takes to replace key components, reliability and energy consumption. The parameters of robots coming from different manufacturers are becoming more and more similar in the upcoming generations, but one of the elements differentiating those products is the software and the costs of use. Rising energy costs are making energy efficiency as important as the payload capacity and accuracy. The growing importance of energy efficiency is reflected by the design of new robots.

The Kyoto Protocol was signed by 192 countries and has resulted in many economies taking a closer look at the level of CO2 emission and the ways of reducing it. Carbon dioxide is mainly produced by power plants, steelworks, chemical industry – branches of the economy that are not robotised very well. The automotive industry, although also energy-intensive, has less of an impact on greenhouse gas emissions. However, robots make up a noticeable part of the energy consumption during the car production process. In 2016, Daimler reported that an average of 2.99 MWh was needed to produce one passenger car. Ten years earlier, in 2006, it was 4.66 MWh [17]. Rising electricity prices cause the robots' energy efficiency to become more and more important – it certainly is a bargaining chip for long-term investments in large-scale robotics.

The data on energy consumption in automotive plants is very interesting. The most robotised lines are the ones producing bodywork (*Body shop*). Robots in a typical car factory are responsible for around 5–10% of energy consumption, so there is definitely something to fight for. Comprehensive research on the energy consumption of robots was carried out at the Mercedes plant in Sindelfingen.

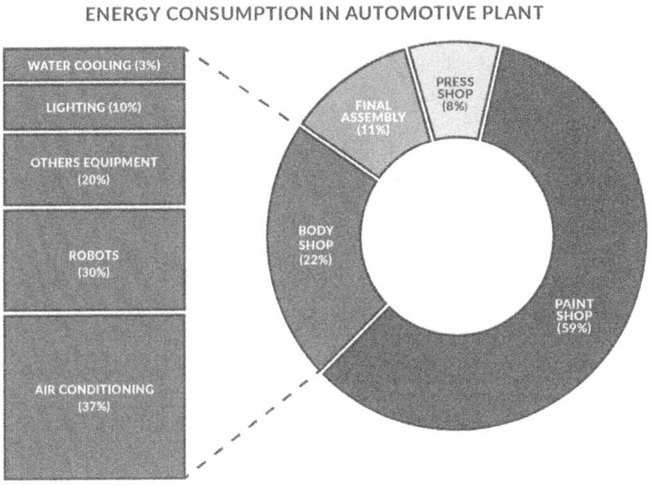

Fig. 36: Energy consumption of the car plant [9].

Statistics concerning robot's work appear to be very interesting. It seems natural for manipulators to be placed in factories to work continuously. The machine, unlike humans, is not getting tired and delivers the same quality at ten o'clock in the morning and three o'clock at night. It would seem logical that it should work all the time so that the investment pays off as soon as possible. However, practice shows that robots do nothing for a long time... (if we understand no movement as "nothing").

Data collected on the bodywork production line shows that the robot is in motion only for 19% of the total time during the year (downtimes longer than 2 seconds were taken as an inactivity). Idle time is caused mostly by days off work, holidays and maintenance stops [18]. During the production process, robots wait about 10% of their time for the motion to be allowed (see chapter 6.1) or the number of the next program to be executed.

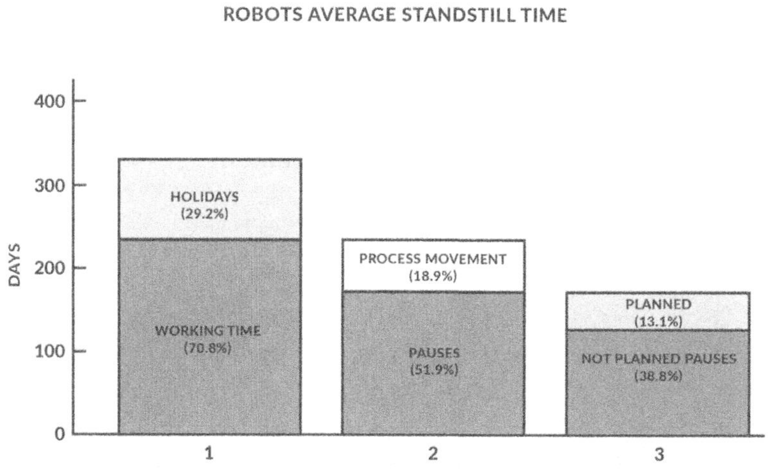

Fig. 37: Average robot's working time per year.

In order to reduce the costs of electrical energy robots are built from lighter materials (less energy is needed for accelerating the robot and lost for stopping it). Manufacturers also use transmissions with reduced frictions, they use mechanical brakes more often and they choose more cost-effective drives. Additionally, reducing the weight of the robot-mounted tools (and shifting their centre of gravity closer to the tool flange) is more energy-efficient.

Setting the robot in motion consumes the most energy (2.5-3.5 kW), while keeping it in one position requires much less energy. When the brakes on the axes are not activated, the robot uses approx. 0.65-0.8 kW (depending on the robot's position and its payload). When the robot is held in the position by brakes, approximately 0.2 kW of energy is required.

Every industrial robot uses brakes for which the locked position is default. Keeping them open requires approx. 100-130 W of energy. After the robot stops, its position is maintained for some time (depending on the manufacturer's settings) by the servomotors. After a long period of inactivity, the brakes are released and are then responsible for keeping the robot in a given position.

One could say that we have found a way to easily increase the energy efficiency of robots. Since the manipulator performs a significant part of its work while standing still, then shortening the time of transition to the stationary state can easily reduce the cost of electricity bills. However, that makes service life of the brakes becomes a problem here. Reducing the waiting time to 2 seconds in case of a robot working on a bodywork production line in the automotive industry results in up to 9.5 million braking cycles in fourteen years of operation. If we take a more conservative approach (20 seconds), we get 1.7 million brakes. At 5 seconds we have less than 5 million braking cycles. Aggressive setting of transition time into the stopped state can significantly reduce brakes' life. As a result, service costs can exceed energy savings, and there is a greater risk of unexpected failure.

According to *Nikkei Asian* Review, an unplanned shutdown in a car plant costs $16,900 per minute [19]. Therefore, when designing a robotic production line, a great deal of attention is paid to the possibility of replacing defective devices in a short time. In car factories, a complete replacement of a robot cannot last more than 60 minutes, it should not take more than 30 minutes to change the gripper and 10 minutes, at the most, to replace the water supply and compressed air hoses.

4

INDUSTRIAL ROBOT'S MOTION

In order to start controlling a robot properly, you need to master the basic theory first. In this chapter you will learn about coordinate systems, motion types and their parameters.

A huge progress in speed, precision and available motion types occurred since the Unimate 1900 was introduced to the market. Initially, manipulators did not offer interpolated motions, and the first programming languages for robots were not introduced until the 1970s. The possibilities and the quality of a robot's motion are highly dependent on the controller. Mechanical units (drives, transmissions, cable routing, materials used) also evolve, but their development is not as vivid as in case of controllers.

Many problems described in this chapter are connected to one another, and the perfect solution would be to explain them all at once. Unfortunately, this is not possible. This is why I strongly encourage you to browse through the chapters and sub-chapters for more information. If you happen to find something you do not understand in one of those – feel free to move to the section devoted to this problem.

Robot programming process usually requires moving the robot into a specified location in the space and remembering its position. The next step is to optimise the motion parameters and to add the logic required to control the external devices. This is how the new robot's motion trajectory is being created, step by step. Execution of the arms motion can occur in two ways: in the joint coordinate system or the Cartesian coordinate system.

4.1. Joint coordinates

It is possible to manually control the angular position of every servomotor in every industrial robot. By changing the angle of rotation in a negative or positive direction, you are able to reach any point in the manipulator's workspace. Respective axes of a mechanical unit are marked with the rotation direction, and the control panel usually has dedicated keys (a joystick in case of ABB's robots), and every key is labelled with the number of axis it controls.

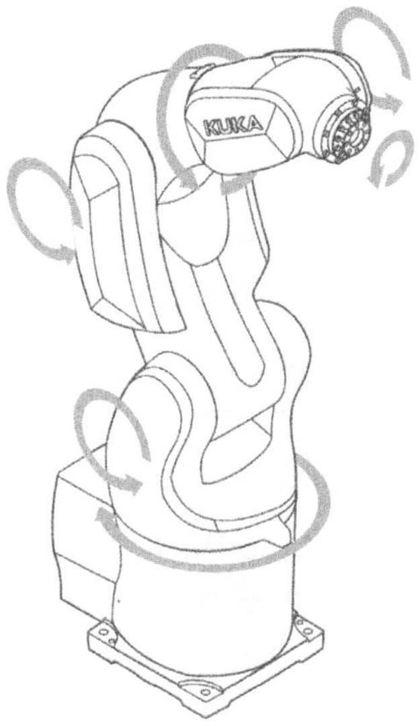

Fig. 38: Joint coordinates.

In order to perform the "obvious" motion (manipulator's arm always moves in accordance with the direction of arrows on it) we need to pay the price – establishing the final position becomes more difficult. If you need to move the end of a tool mounted on the robot by 20 cm to the left, the only possible way is through trial and error. No one is able to calculate in their head what would be the required rotation of respective axes in order to achieve the desired displacement while maintaining the tool's orientation. What would be the displacement of a tool mounted on a robot if we modify the rotation angle by 2° on the second axis and 3.5° on the third? And what will happen if we additionally change the location of the fifth axis? After some time and acquiring some experience, it turns out that in majority of cases you can easily achieve an approximate robot's position in the workspace by simply controlling the angles of servomotors' rotation. However, any attempts of precise offset, by specified values, while maintaining the tool's orientation, are bound to fail.

Moving from the start position to the first processing point with the utilisation of the joint coordinates is an excellent idea. We do not require movement on a strict trajectory here, but we want it to be collision-free. Control with the utilisation of joint coordinate system usually involves only the required number of drives, allowing the movement to be faster and smoother compared to motion with the use of a Cartesian coordinate system.

As you have already learned in chapter 3.2, respective robot drives have different maximum velocities. Soon you will learn that when the program is executed, the fastest axis of the robot needs to adjust its work speed to the slowest one that's taking part in a given motion. Limiting the number of drives involved in a robot's motion is important if we care about time-optimal trajectory. If we need to move the robot somewhere, why shouldn't we use the fastest axes? Intuition tells us that we should use drives no. 4, 5 or 6, because, according to the catalogue, they can develop the highest angular velocity. In practice, especially in case of big robots, a better choice is axis no. 1 or 2, because even a small movement of those axes results in a big change of kinematic chains position, especially when it is extended.

Apart from executing general motions, joint coordinates are also used when we need to set the robot's arms in a calibration position and when it is required to move through a singularity configuration (described in chapter 4.6).

By saving the program in the joint coordinates, we save the position of the arms (angle of rotation for all robot's drives) by using special commands in the robot programming languages. The position is defined by establishing the rotation angle for specific axes. Points defined in such a way do not keep the record of the robot's configuration (for more information see sub-chapter 4.5), because it is explicit and it results from the values of drives rotations.

Syntax example: PointName (A1, A2, A3, A4, A5, A6)

ABB example:

```
CONST jointtarget HOME:=[[0,-50,30,-90,-75,115],
[9e+09,9e+09,9e+09,9e+09,9e+09,9e+09]]
```

KUKA example:

```
DECL E6AXIS XHOME={A1 0,A2 -90,A3 90,A4 0,A5 0,A6 0}
```

Fanuc example:

```
P[2] {
 GP1 :
 UF : 0, UT : 0,
 J1 = 0.000 deg, J2 = 0.000 deg, J3 = 0.000 deg,
 J4 = 0.000 deg, J5 = -90.000 deg, J6 = 0.000 deg
};
```

4.2. Cartesian coordinates

Steering the robot in a Cartesian coordinate system seems to be the most intuitive operating method for people who hold the control console for the first time. If we want to perform a simple activity, such as moving the robot's end up in a straight line, the only thing we need to do is to change the value of position on the Z axis. Front/back and left/right movements can be attained by changing the position along the X and Y axes of the Cartesian coordinate system. The robot controller automatically calculate offsets for the appropriate angular movements of respective drives.

The only trap that we can encounter (but a very tricky one) is the movement with the use of a different coordinate system than we thought we would. Modern robots offer the opportunity to predefine dozens of supporting reference coordinate systems (more in chapter 4.4). If we need to move within a coordinate system, whose position in space is unknown, then the robot's movement may seem to be a little random, which is frustrating and may complicate your work.

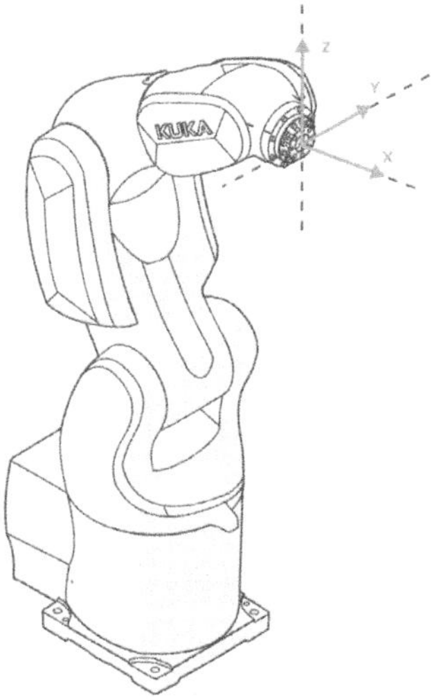

Fig. 39: Cartesian coordinates (XYZ).

Position of some of those coordinate systems is determined based on the data from the assembly documentation (e.g. TCP – subchapter 4.4.2). However, sometimes a mistake can occur during the assembly of a tool on a robot, and the definition entered into the controller does not correspond to reality. This creates a situation in which the coordinate system that was supposed to be helpful, works against us instead. Nevertheless, there is nothing stopping us from defining it in the right way, so that we can enjoy the intuitive robot control once again.

While in case of offsets (X, Y, Z) such mistakes are not a huge inconvenience, in case of rotations in the Cartesian space (Rx, Ry, Rz) with the utilisation of an undefined reference coordinate system can be really troublesome. I will discuss this more in the chapter devoted to coordinate systems.

We are using, by default, one of two standard reference coordinate systems: the one located in the robot's base (base coordinate system) or the one on the tool flange, where the tool is being installed (tool coordinate system). The controller is responsible for calculating robot's end position. It uses the algorithm of inverse kinematics and calculates the angular position of all drives on a current basis in a manner that allows the robot to move in a straight line, in accordance with the axes (X, Y or Z) of the assumed coordinate system.

In order to reach the position given by the operator, the robot usually uses all drives (in order to maintain constant orientation of the tool), which leads to creation of less time-optimal programs. If the programmer used only Cartesian coordinate system during programming, then the robot's movement in the final program is much more "wobbly" and less smooth. If we want to execute a general movement which does not require high precision, then usually the best idea is to manually control the rotation angles of respective axes (joint coordinate system).

Cartesian coordinate system works very well when it is required to make a movement by a specified distance in a straight line and to change the position of the tool while maintaining its orientation. When you have correctly declared coordinate systems, you will be able to significantly speed up programming for such processes as palletizing, welding, gluing and many others.

Recording a robot's position in Cartesian coordinates requires registering configuration of the manipulator's arms, because a standard six-axis robot can reach majority of points in a working space in many ways (chapter 4.5). It is also the reason why additional information is required in order to unequivocally establish position of the robot's arms. By default, the values appearing in the point's definition are the distance from the robot's end (tool flange) to its base.

Syntax example: PointName (X, Y, Z, Rx, Ry, Rz, arms configuration)

ABB example:

```
CONST robtarget P10:=[[2486.99,-192.55,197.08],[-0.000001,1,0,0],
[0,-1,0,1],[9e+09,9e+09,9e+09,9e+09,9e+09,9e+09]]
```

KUKA example:

```
DECL E6POS XP10={X 445,Y 0,Z 810,A 180,B 0,C 180,S 2,T 10}
```

Fanuc example:

```
P[1] {
 GP1 :
 UF : 0, UT : 0, NUT 000
 X = 440.000 mm, Y = 0.000 mm, Z = 245.000 mm,
 W = 180.000 deg, P = 0.000 deg, R = 0.000 deg
};
```

4.3. Establishing robot's position

Robot's control system contains information regarding mathematical beginning of the mechanical unit (robot's base) and knows lengths of respective arms and angular position of all servomotors. The controller, depending on the selected controlling mode (joint coordinate system or Cartesian coordinate system) needs to solve one of two equations in order for the mechanical unit's movement to be possible. These equations are related to forward kinematics (joint coordinate system) or inverse kinematics problems (Cartesian coordinate system).

Tab. 11: Methods of establishing robot's position.

Type of task	Data	Unknown variables
Forward kinematics	Arms' lengths rotation angles of drives (α, β, λ)	position in space (X, Y, Z)
Inverse kinematics	Arms' lengths position in space (X, Y, Z)	angular position of drives (α, β, λ)

It is not required for a robot programmer to be familiar with the course of those calculations, unless you find a job in a company that writes programs for robot's control cabinets. But in that case, you would need much more mathematics than this book offers.

I am mentioning this, because I have encountered this common belief many times, that robot programming is basically creating everything from scratch. It sounds strange to me, because when writing the first Hello World! program for your PC or smartphone, no one is assuming that it will be required to manually create operational system and drivers for the equipment as well. This is similar with robots – robot's controller is already equipped with a command interpreter and all the software needed to create and execute programs.

The understanding of forward and inverse kinematics' problems helps to comprehend what processes are taking place inside the robot, and how to create better programs. It also allows us to be aware of some things, e.g. why is it necessary to save the configuration of arms when moving in Cartesian coordinate system, and where do the singularities come from (chapter 4.6).

Forward kinematics allows us to estimate what will be the change of kinematic chain's end position in case of a rotation angle for any of robot's axes changes. The result of those calculations is unambiguous and only one correct solution is possible (Fig. 40a).

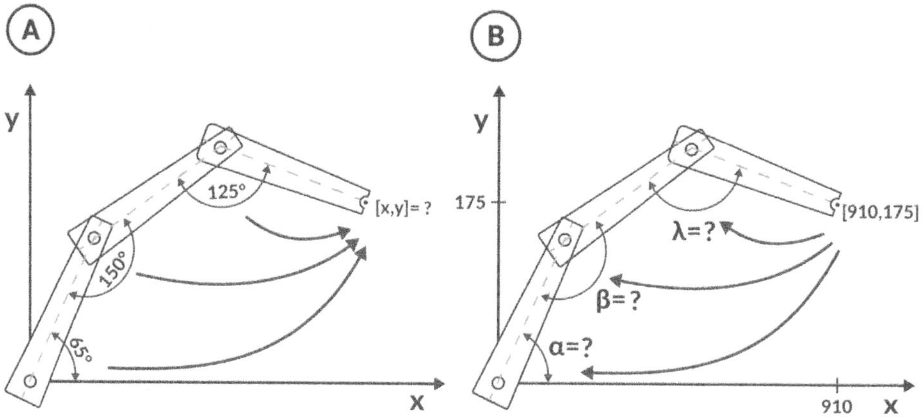

Fig. 40: Task related to a) forward kinematics, b) inverse kinematics

Inverse kinematics approaches this matter from a different perspective. In this case, we check the angles of rotation that each axis must have, if the end of our robot is at a known point in the work space (Fig. 40b).

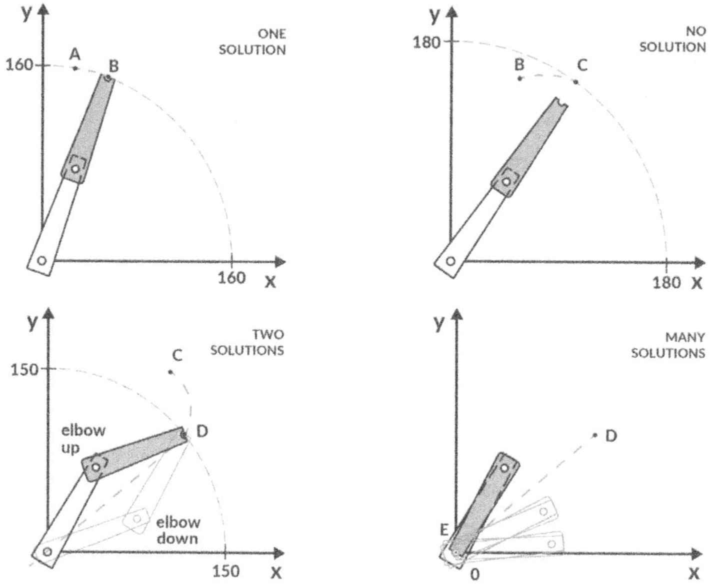

Fig. 41: Inverse kinematics – possible solutions depending on the location of a target point in space.

The process in case of inverse kinematics is much more complicated. The number of solutions may vary from case to case. It may turn out that there is more than one solution, there are infinitely many solutions or not even one correct result (Fig. 41). The calculations become more complicated as the number of robot axes increases.

Usually there is more than one correct solution, and that results in serious consequences. The position of the robot in space cannot be described solely by the position of the robots end. It is also necessary to add information about the arm configuration so that the robot reaches the set position every time in the same way.

4.4. Coordinate systems

If we want to move the robot in Cartesian coordinates, we need reference points that will allow us to unequivocally define our position in space. Fig. 42 shows a very simplified robot diagram in a 2D projection. The robot's task is to move from point A (its current location) to point B. Point B is moved away from the robot's base by X: 150 mm and Y: 60 mm. The robot is able to determine its current position on the basis of arm lengths and angles of rotation of individual drives (forward kinematics task). It is equal to X: 140 mm Y: 80 mm. In order to get to point B with known coordinates (150/60), the robot needs to calculate how much it needs to turn the individual drives to get to the desired location. It is done with the use of inverse kinematics.

If the next steps of the robot controller's work were to be presented, they would look like this:

- calculating the current Cartesian position of the robot (using forward kinematics) based on the angular position of its drives and known arm length
- checking if the current position does not coincide with the target position

- calculating the target drive positions based on inverse kinematics (and verification if they are reachable)
- rotation of the drives by the differences between the current and target rotational values of the drives.

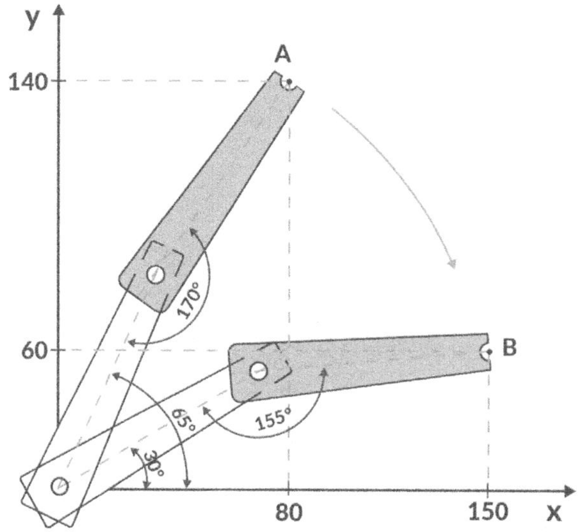

Fig. 42: Robot's motion.

As it can be seen, in the simplest case, the only data the robot needs to know is the distance of the next points from its base, the points through which it has to pass.

It would also be possible to create a program in which all the subsequent path points are stored in a joint coordinate system. Then the controller's work would be limited to reading these values and rotating the servomotors in accordance with them (but while observing certain additional rules – see subchapter 4.7.1).

In most cases, however, the positions of the subsequent points in the robot programs are recorded using the Cartesian coordinate system. Such recording convention provides a number of additional possibilities, including easier adaptation of offline programs to the actual situation on the production line.

All the above principles remain valid in case of a real robot located in 3D space. However, additional coordinate systems are being introduced in order to simplify the operation and allow the robot to be programmed more quickly, while reducing the number of possible errors. We can significantly simplify our work by using additional coordinate systems.

We use the following reference systems in industrial robotics:

- Base Coordinate System
- Tool Coordinate System
- User Coordinate System

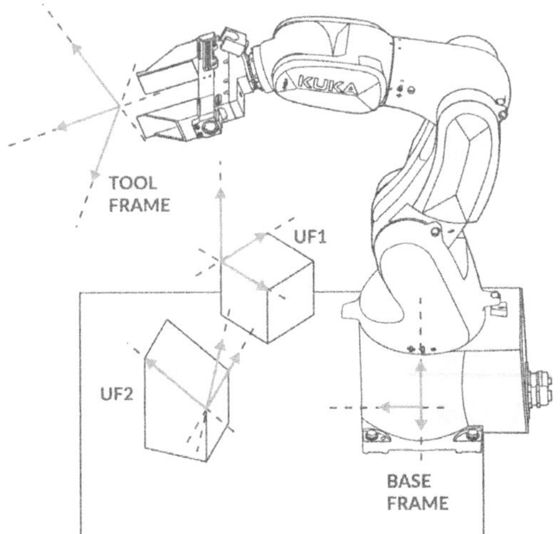

Fig. 43: Robot coordinate systems.

All these systems and their positions are shown in Fig. 43. One robot and three types of coordinate systems? Fortunately, it is a bit simpler in practice. The robot needs only two of them to move (tool and base coordinate systems), and they are defined by the manufacturer in each manipulator by default. However, it is worth adjusting them for the task being performed, as attempts to oversimplify the coordinate systems in-use usually complicates the work.

So let's have a look at our reference systems. The easiest way to do this is to analyse parts of the actual robot programs. I will use two identical examples from ABB RAPID (IRB 140 robot) and KUKA KRL (Agilus KR6 R900 sixx model) language. Thanks to that, you will be able to see that despite the completely different syntaxes of these programs (each robot manufacturer has its own programming language that's compatible only with its own products), all industrial robots operate according to the same rules.

Let's create the first, simplest program, which will aim to achieve position defined by us. First of all, we will save the program in the joint coordinates for the robots from Fig. 44.

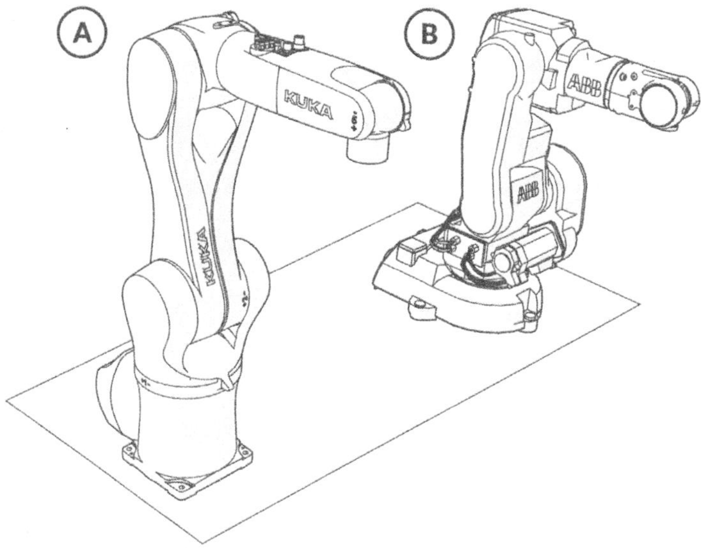

Fig. 44: KUKA Agilus KR6 R900 sixx and ABB IRB 140.

ABB:

```
LOCAL CONST jointtarget HOME:=[[0,0,0,0,90,0]];
MoveAbsJ HOME,vmax,fine,tool0\Wobj:=wobj0;
```

KUKA:

```
DECL E6AXIS HOME={A1 0,A2 -90,A3 90,A4 0,A5 90,A6 0}
PTP HOME Vel=100 % PHOME
```

Although the robots are arranged in the same way, the values for the angles of rotation on the axes are not the same:

Robot's axis:	A1	A2	A3	A4	A5	A6
ABB IRB 140	0°	0°	0°	0°	90°	0°
KUKA KR6 R900 sixx	0°	-90°	90°	0°	90°	0°

These differences are the result of different conventions adopted by the robot manufacturers, other position of the servomotor shaft on A2 and A3 axes in ABB and KUKA are treated as zero value.

KUKA robot stores its program in two files, and I have decided not to include the complete version of it for sake of legibility of the list. The *PHOME* constant in this case looks as follows:

```
DECL E6AXIS HOME={A1 0,A2 -90,A3 90,A4 0,A5 0,A6 0}

DECL FDAT FHOME={TOOL_NO 0,BASE_NO 0,IPO_FRAME
#BASE,POINT2[] " ",TQ_STATE FALSE}

DECL PDAT PPHOME={VEL 100,ACC 100,APO_DIST 0,
GEAR_JERK 50}
```

From this one can derive the following information – the velocity and acceleration is 100% (*VEL 100,ACC 100*) and the robot reaches the point with full accuracy (*APO_DIST 0*).

To make it easier to understand, I have also prepared some simplifications to the ABB program code. The full declaration of the HOME variable looks like this:

```
LOCAL CONST jointtarget HOME:=[[0,0,0,0,90,0],
[9e+09,9e+09,9e+09,9e+09,9e+09,9e+09]];
```

Record `9e+09` means that no additional robot axes have been defined (e.g. linear motion track, Fig. 18) and the robot controller ignores it.

Apart from the recording method, the rest is the same. Robot's starting position is defined by a constant called HOME, which stores the rotation values of the drives expressed in degrees. Then there is a motion instruction (PTP / MoveAbsJ), which tells the controller how to rotate the drives to reach the HOME point (for more information see chapter 4.7). Additionally, we also have declared parameters of this movement, such as velocity and accuracy (ABB: *vmax, fine*).

These parameters are also provided in case of the KUKA robot, but they are saved as a reference with another constant: *PHOME.*

Robots following the above instructions will always reach the position defined in the HOME constant. The above programs contain information on coordinate systems:

- tools: tool0 (ABB) / TOOL_NO 0 (KUKA)
- robot base: wobj0 (ABB) / BASE_NO 0 (KUKA).

They are ignored by the controller because we have saved the robot's position in joint coordinates, which is indicated by the type of variable: *jointtarget* (ABB) and E6AXIS (KUKA).

4.4.1. Robot base system

Let's try to convert the program from the previous subchapter from joint coordinates to Cartesian coordinates. We will need two coordinate systems for this: robot base coordinate system and tool coordinate system. Let us start by explaining the nature of the first one.

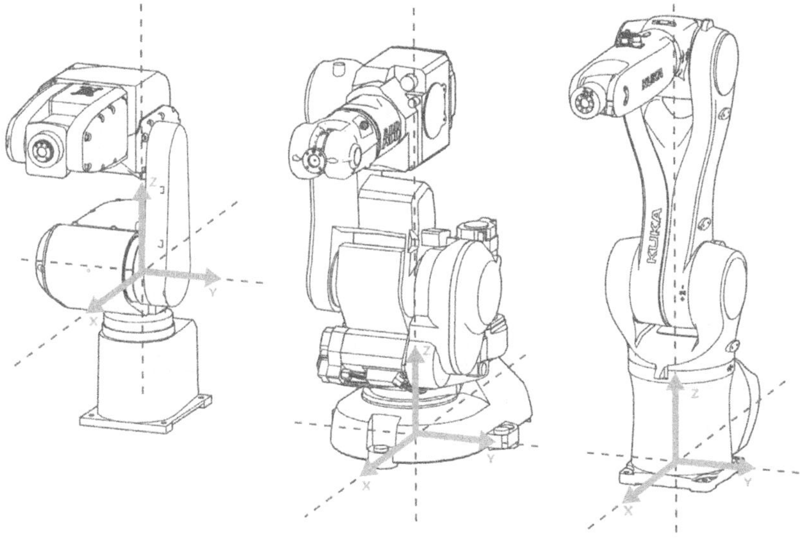

Fig. 45: Location of the base in different robots.

The zero point of this coordinate system is located in the robot's base. Fig. 45 shows examples of where this reference system is located in several different robots from different manufacturers: Fanuc LR Mate 200iB, KUKA KR6 R900 sixx, ABB IRB 140. As you can see, it can be located at different heights, but it is always in the axis of the robot's base and it is a fixed, unchangeable position declared by the manufacturer for a given model.

It can be said that the robot's base coordinate system is the "beginning" of a mechanical unit. It is a fixed system that is "mounted" permanently to the robot's base. If the programmer does not define his own part coordinate system (chapter 4.4.3), the position of the robot end is calculated with reference to the robot's base coordinate system which is inflexible and cannot be moved (unless the whole robot is moved with it). Therefore, part coordinate system is usually used to program the robots. The robot's base coordinate system is typically used to create programs for simple applications.

A block of code from the previous section in the RAPID language for the ABB robot looks as follows:

```
!******** BASE DATA ********

PERS wobjdata base0:=
[False,True,"",[[0,0,0],[1,0,0,0]],[[0,0,0],[1,0,0,0]]];

!******** TOOL DATA ********

PERS tooldata tool0:=[True, [[0,0,0],[1,0,0,0]],
[0.001, [0,0,0],[1,0,0,0], 0,0,0]];

! ******** ROBTARGET********

CONST robtarget P20:=[[450,0,647],[0,0,1,0],[0,0,0,0],
[9e+09,9e+09,9e+09,9e+09,9e+09,9e+09]];

MoveJ P20,vmax,fine,tool0\Wobj:=base0;
```

Comments in the program start with an exclamation mark (!). The BASE DATA and TOOL DATA sections contain declarations of the used coordinate systems. This chapter focuses on the base coordinate system (*wobjdata* constant).

Then there is the declaration of the newly defined *P20* point, which is a record of the position in the Cartesian space. The position of the robot at this point is identical to that of *HOME* point from the previous chapter (joint coordinates A1: 0°, A2: 0°, A3: 0°, A4: 0°, A5: 90°, A6: 0°). In addition to the name, type of the constant has changed as well – now it is r*obtarget* instead of a *jointtarget.* That means the robot's controller knows it is going to interpret this line differently (i.e. as a Cartesian record, not as a joint record).

Finally, there is the motion instruction (*MoveJ*), which defines which point (*P20*) the robot should reach using which coordinate systems (tool0/base0) and what velocity (*vmax*) and accuracy (*fine*) parameters. It looks the same as one in the case of the program written in joint coordinates.

The KRL language being used by the KUKA robot has a different syntax than ABB RAPID (additionally, the declaration of coordinate systems, the coordinates of program points and motion parameters are broken down into different files), but we should be able to find common elements without any problems:

```
;******** BASE DATA ********

BASE_DATA[0]={X 0,Y 0,Z 0,A 0,B 0,C 0}
BASE_NAME[0,]="Base0"

;******** TOOL DATA ********

TOOL_DATA[0]={X 0,Y 0,Z 0,A 0,B 0,C 0}
TOOL_NAME[0,]="Tool0"

DECL E6POS XP20={X 445,Y 0,Z 810,A 180,B 0,C 180,S 2,T 10}

PTP P20 Vel=100 % PP20 Tool[0]:Tool0 Base[0]:Base0
```

We start from the coordinates system declaration, and the robot's position is presented below. Type has changed here as well (from *E6AXIS* to *E6POS*), so the robot knows that the Cartesian coordinate system would be used to calculate robot position.

Finally, there is the *PTP* motion instruction (equivalent to *MoveJ* in ABB RAPID). As in the first example, we have the definition of the position (in KRL language it is preceded by the X character), reference to the point *P20*, maximum velocity of motion (*Vel=100%*) and the used coordinate systems *(Tool[0]*, *Base[0])*. Comments are preceded by a semicolon (;).

Let's have a look at the practical use of the robot's base coordinate system. It is important to know how the robot records its position in the Cartesian reference coordinate system. Let's go back to the first list that has appeared in this chapter and look at the declaration of the *P20* point, written in the ABB RAPID language:

```
CONST robtarget P20:=[[450,0,647],[0,0,-1,0],[0,0,0,0],
[9e+09,9e+09,9e+09,9e+09,9e+09,9e+09]];
```

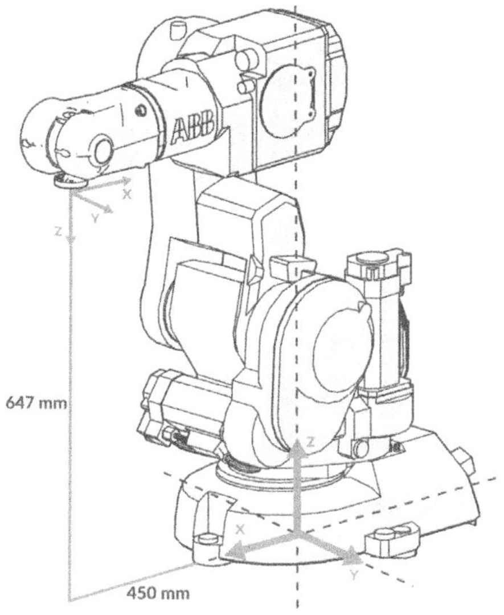

Fig. 46: Robot's position recorded in Cartesian coordinates.

Values [*450,0,647*] correspond to a 450 mm offset in the X axis, 0 mm in the Y axis and 647 mm in the Z axis, and [*0,0,-1,0*] means that orientation is stored in a slightly unusual way, hence quaternions are used instead of Euler angles. Value [*0,0,-1,0*] corresponds to the values Rx: 180°, Ry: 0°, Rz: 180°. The second to last square bracket [*0,0,0,0*] is a record of the arms position. Values [9e+09,9e+09,9e+09,9e+09,9e+09] inform us that the robot does not have any additional axes defined (e.g. the linear motion track).

Therefore, we have all the data we need to determine the position of the robot's end. So far we have not defined any of our own coordinate systems, but we use what has been saved in the controller by the manufacturer instead. The robot will attempt to place the tool flange on the 6th axis at a distance of [*450,0,647*] from its base. It can be seen in Fig. 46.

For the KUKA robot, record of the axial position HOME (A1: 0°, A2: -90°, A3: 90°, A4: 0°, A5: 90°, A6: 0°) in the Cartesian coordinates (as *XP20* constant) shall follow the same rules. We use the default coordinate systems of the robot's tool and base here as well.

```
DECL E6POS XP20={X 445,Y 0,Z 810,A 180,B 0,C 180,S 2,T 10}
```

The information about the arms configuration in KRL language is more eye-catching: *S 2.T 10*, just to remind, in case of ABB it is [*0,0,0,0*]. Unfortunately, both records are unclear for people. Chapter 4.5 tells more about why it is important to write down the position of the arms in the robot and what it actually means.

Similarly to creating records in joint coordinates, you may notice that even in the Cartesian system, the coordinates that describe the same arrangement of robot arms are not identical:

Robot name	X [mm]	Y [mm]	Z [mm]
ABB IRB 140	450	0	647
KUKA KR6 R900 sixx	455	0	810

It is the result of mechanical design differences. Both robots have different arm lengths, and their range is not the same either. Y-axis is equal to zero because the end of both robots lies on the same axis with the centre of their base, but the height above the floor (Z-axis) and the offset in the X-axis are different.

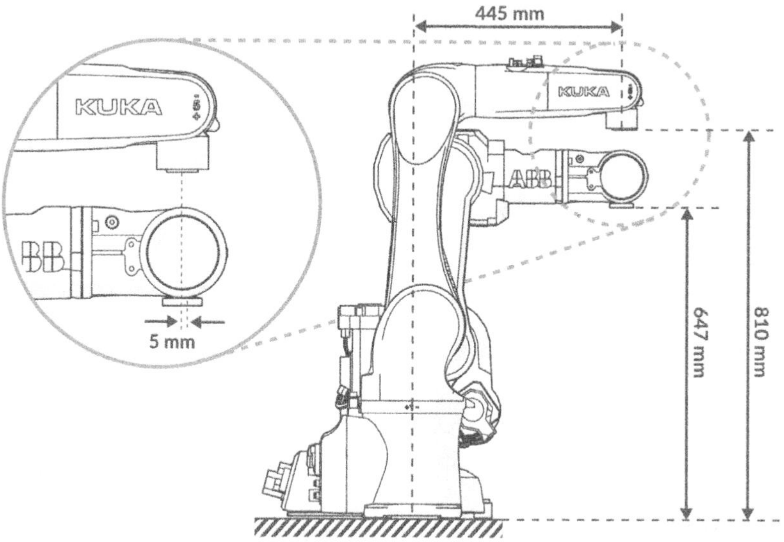

Fig. 47: Differences between KUKA and ABB robot designs.

Once again, I would like to draw attention to the differences in the coordinates. The robot program is always linked to the construction of a mechanical unit. It shall not be transferable between different models of the same manufacturer. If you had two robots, KUKA KR6 R900 (901 mm range) and KR6 R700 (706.7 mm range), then uploading the same program will cause the robot's arms to align a little differently, because of the difference in their length. It is highly probable that part of the program will not be possible to be executed on the KR6 R700, because the robot will not have sufficient arm range to reach the declared points.

4.2.2. Tool coordinate system

I mentioned before that you need two reference coordinate systems to determine the position and orientation of the robot in the Cartesian space. You already have some basic information about the first one – the robot base. The second is the tool coordinate system. I have already shown in the previous subchapter that by using both these systems it is possible to clearly and legibly record the robot's position. No additional coordinate systems would be needed assuming that robot programming was only about recording subsequent positions; it would probably be enough to use only the joint coordinates and to record the angular position of the drives.

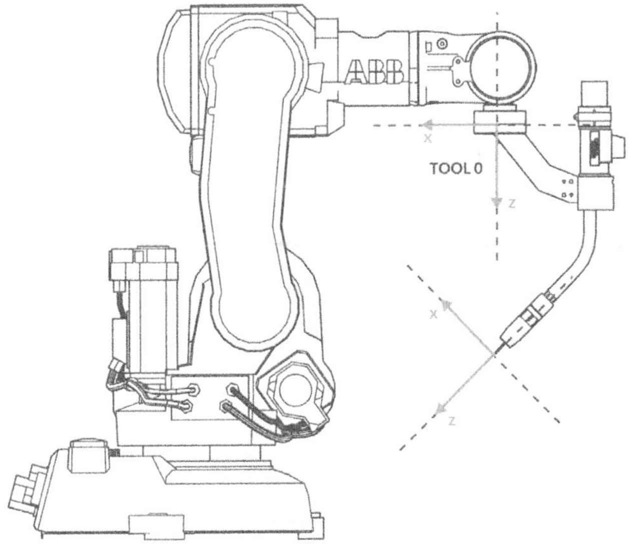

Fig. 48: Robot with mounted tool.

Since programming of a robot requires moving it to a specific position in space, we also care about the intuitiveness of controlling the robot.

A robot that leaves the factory has basic knowledge of its dimensions. We have already talked about the beginning of the kinematic chain and we know that it is defined in the base. For an "unboxed" robot the end of the kinematic chain is defined at the centre of the tool flange on the last axis of the robot.

The situation looks as shown in Fig. 48 after the tool has been mounted. The information about the end of the kinematic chain is no longer up to date, because the robot's length has been extended by the length of the welding torch. While there is nothing to stop you from using default tool coordinate system (called the "zero" tool), it would be, in most cases, inconvenient. Imagine that you need to move along the "Z" axis of the welding wire by 45 mm. It is difficult to execute a linear motion using the "zero" coordinate system of the tool, due to the angular shape of the torch. This requires simultaneous correction of the robot end position along X and Z axes.

This cannot be done intuitively and unambiguously if there is only the default coordinate system defined on the robot's tool flange. Moreover, trying to use joint coordinates (to control the angle of rotation of individual drives) will not make the task any easier. In order to simplify the programming, you would need to define an additional system of coordinates located at the end of the torch.

Before we go further, I would like to say a few words about the – TCP (Tool Centre Point). Each robot-mounted tool has a unique characteristic point relevant to the process it performs. In case of welding, it will be the electrode tip on the fixed welding gun's arm, for arc welding it will be the end of the welding wire, for handling parts it will be the fixed pin on the gripper (which determines the position of the element), and for gluing – the nozzle's end (Fig. 49).

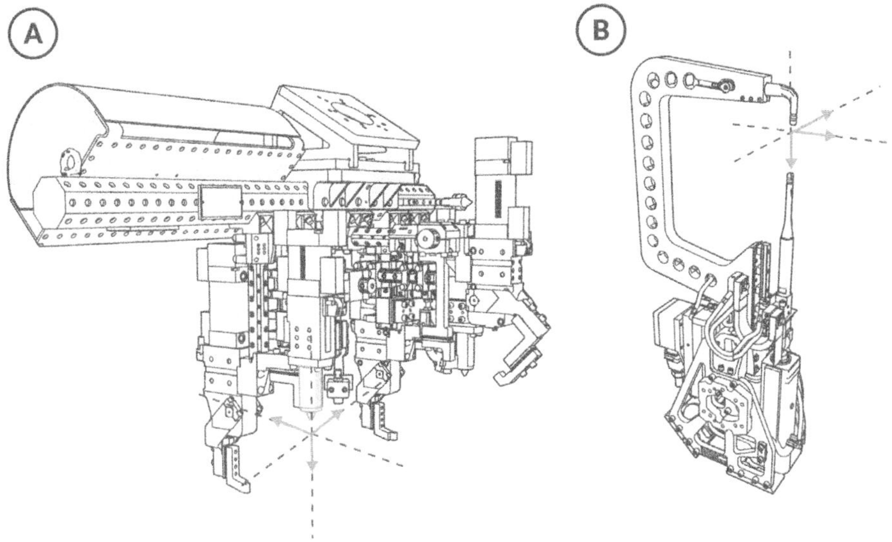

Fig. 49: Characteristic point of the tool on the example of a) gripper, b) welding torch.

The TCP informs us about the position of this point in relation to the mounting of the tool. As an example, I will use K2647-4 Magnum® PRO Robotic 550 welding torch from the Lincoln Electric. Its TCP point is shifted in the X axis by 7 mm, it is in one line with the Y-axis (0 mm), and in the Z-axis the shift is equal to 352 mm. The nozzle has a 45 degree angle (Fig. 50).

The TCP point declaration (which I have called t_Torch) for this MIG/MAG welding torch looks as follows in the ABB RAPID language:

```
!******** TOOL DATA ********

PERS tooldata t_Torch:=[True, [[-7,0,352],[0.92388,0,0.382683,0]],
[0.001, [0,0,0],[1,0,0,0], 0,0,0]];
```

The *True* parameter informs us that the tool is mounted on the robot (so it is not an external tool – see the end of this section for more information). The next information concerns the distance between the end of the welding wire and the tool flange [*-7,0,352*]. Information on the nozzle's angle (45°) is stored in the following quaternions: [*0,0.190809,0,0.981627*]. The following items should be recorded in the next positions: mass of the tool, its centre of gravity and the moments of inertia. I have left them empty for the time being to simplify the code analysis.

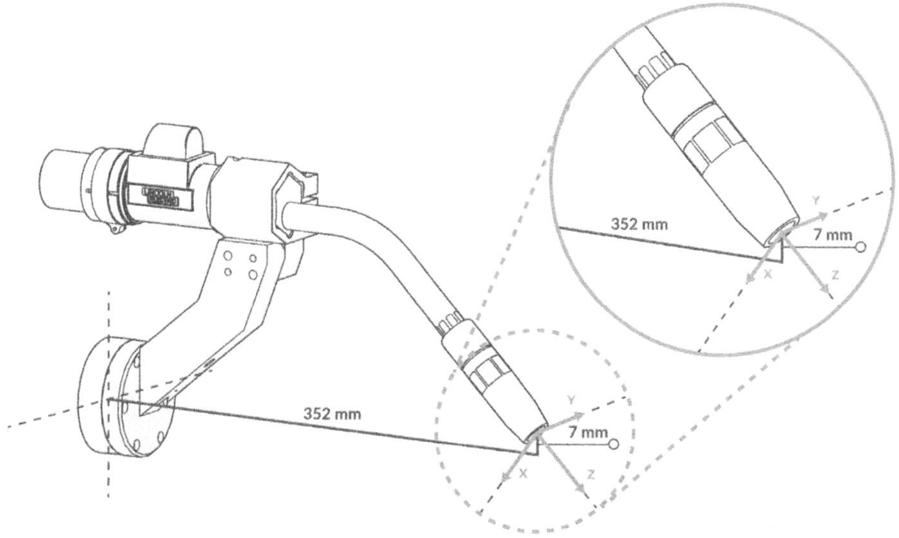

Fig. 50: TCP in the MIG/MAG burner.

Definition of the same tool in the KRL language will look as follows:

```
;******** TOOL DATA ********

TOOL_DATA[1]={X -7,Y 0,Z 352,A 0,B 45,C 0}

TOOL_TYPE[1]=#TCP

TOOL_NAME[1,]="t_Torch"

LOAD_DATA[1]={M -1.00000,CM {X 0.0,Y 0.0,Z 0.0,A 0.0,B 0.0,
C 0.0},J {X 0.0,Y 0.0,Z 0.0}}
```

As you can see, KUKA keeps the record of angles in a more legible manner (A 0,B 45,C 0), but the whole record is divided into four lines, where one contains coordinates, the second is about the type, the third is about the name, and the last about the mass and centres of gravity.

Among interesting differences, it is worth mentioning that ABB recognises subsequent coordinate systems of the tool by a unique name (in this case *t_Torch*), KUKA, on the contrary, by another index (in this case the first one – *TOOL_DATA[1]*). For the KUKA robot, the name is optional, it is used only for the convenience of the programmer. We can see that in both cases the declared values are the same, because the welding torch component itself is identical.

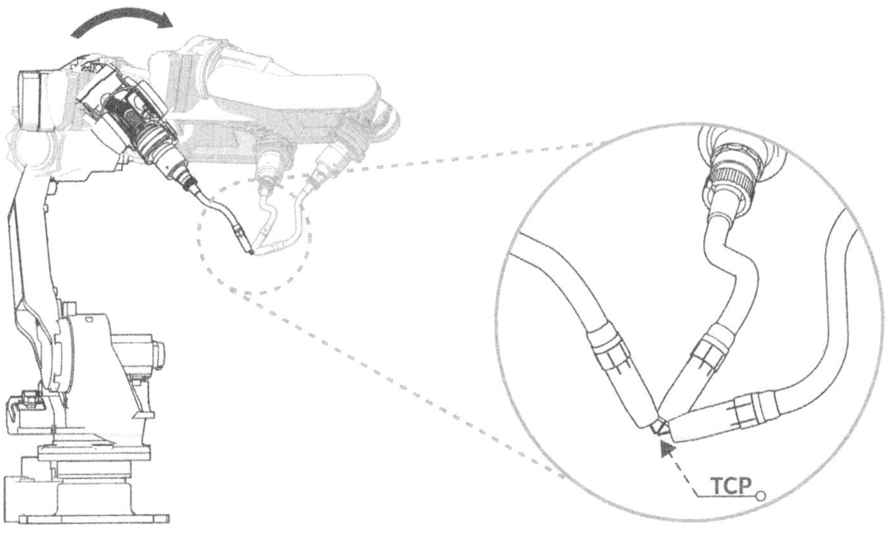

Fig. 51: Verification of the TCP point correctness by rotating around it.

Information about the position of the TCP is obtained from the design documentation (it should also contain information about the mass, centre of gravity and moments of inertia). In case the TCP information is not provided with the tool, it shall be determined manually with the use of methods made available by the robot controller. Verifying the correctness of the declared TCP is very simple and consists of turning the robot in relation to TCP in the axes Rx, Ry, Rz. During these operations, the orientation (in this case) of the welding torch should change, but the welding wire located at its end will be at a fixed point in the space (Fig. 51). In case it is not, check whether you have selected the correct coordinate system on the control panel or enter/determine the TCP again. The robot may have multiple tool coordinate systems defined for different tools, but only one can be active at a time.

If the operator wants to move the tool in a straight line while maintaining its orientation then using the tool coordinate system is the easiest solution. Additionally the robot moves along the X, Y and Z axes which are defined at the tip of the tool.

Having defined the coordinate system for the tool, we can modify the program to use it.

```
!******** TOOL DATA ********

PERS tooldata t_Torch:= [True, [[-7,0,352],[0.92388,0,0.382683,0]],
[0.001, [0,0,0],[1,0,0,0], 0,0,0]];

CONST robtarget P20:=[[457,0,295],[0.382683,0,-0.923879,0],
[-1,-1,-1,6],[9e+09,9e+09,9e+09,9e+09,9e+09,9e+09]];

MoveJ P20,vmax,fine,t_Torch\Wobj:=wobj0;
```

Let's start the analysis from the end in order to see what has changed. In the last line (in the MoveJ motion instruction) we are using a new tool coordinate system t_Torch (instead tool0), which we have declared a few lines before. The changes affected the values of the X, Y, Z coordinates at P20 point, and instead of *[450,0,647]*, we have [*457,0,295*] now. The declared angles of rotation have changed in the same way. The robot now considers the welding torch's nozzle to be the end of its arm and calculates the coordinates for it in the Cartesian coordinate system. If you attempt to reach a modified P20 point using a new tool coordinate system, the robot will not change its position (Fig. 52A).

Let's go back for a moment to the "zero" tool (*tool0*), leaving the revised P20 coordinates used for the t_Torch tool unchanged:

```
CONST robtarget P20:=[[457,0,295],[0.382683,0,-0.923879,0],
[0,-1,0,6],[9e+09,9e+09,9e+09,9e+09,9e+09,9e+09]];

MoveJ P20,vmax,fine,tool0\Wobj:=base0;
```

If you try to follow these instructions, the tool will come dangerously close to the floor (Fig. 52b). This is not the position that we wanted to achieve. When working with a real robot, such change is very likely (depending on the initial position of the arms, the shape of the mounted tool and the surrounding elements) to lead to collisions and damaging very expensive equipment.

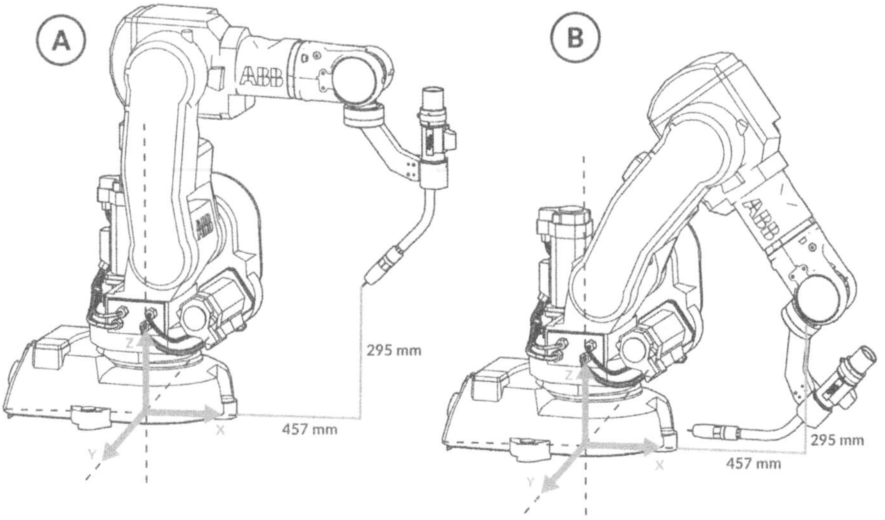

Fig. 52: Effects of changing the tool definition in the program without changing the value of the point coordinates.

As you can see, changing the definition of the tool used in the program is a very serious matter. Doing so manually (by only modifying the information about the used tool without modifying the point coordinates) will make the robot move in a completely different way than before.

Such partial program modifications are unacceptable – they can lead to life threatening situations and/or severe losses. Each point in the program, where the declaration of the tool used has been changed, requires modifying. You can do this manually by reaching a given point, modifying the declaration of the coordinate system used in the tool, and then force the robot to calculate the new coordinates and replacing the previous ones. After such change, try to reach the point that has just been corrected. If the robot does not move, it means that all changes have been made correctly.

In order to ensure that the entire program continues to run properly, test it from start to finish at low motion velocities.

If we perform the same operation to change the tool on the KUKA robot, the program, in the end, will look as follows:

```
!******** TOOL DATA ********

TOOL_DATA[1]={X -7,Y 0,Z 352,A 0,B 45,C 0}
TOOL_TYPE[1]=#BASE
TOOL_NAME[1,]="t_Torch"

DECL E6POS XP20={X 452,Y 0,Z 458,A 180,B -45,C 180,S 2,T 43}
PTP P20 Vel=100 % PP20 Tool[1]:t_Torch Base[0]
```

Individual coordinate systems defined for each tool are such big help in programming. In fact they are almost always used in industrial practice. Programs for the automatic tool changing or specific service routine could be an exception. Tool definitions are used not only for the convenience of the programmer during creation of the program. Their application offers other, additional advantages.

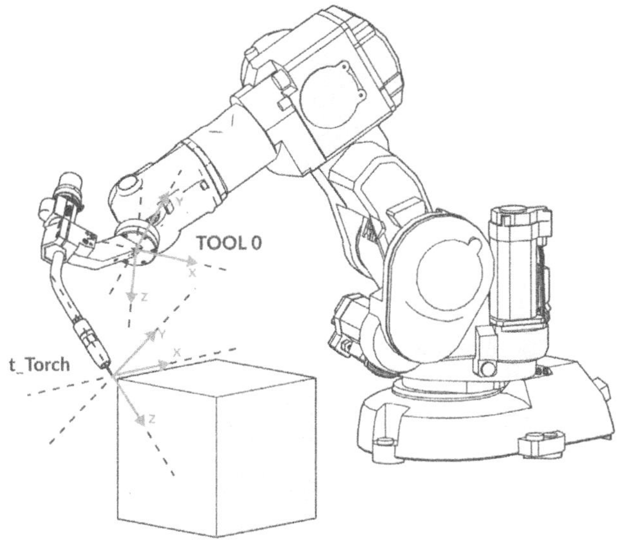

Fig. 53: Programming robot's movement along the edge of the cube is much easier if you use an additional coordinate system.

In arc welding, the welding torch's nozzle is cleaned from time to time, and the welding wire with the TCP defined at its tip is trimmed. Special devices such as ABB Bull's eye or Motoman Q-TCP are used to automatically determine the position of the new TCP. They allow to measure the difference between the previously declared TCP and the situation after the execution of the wire trimming program. The result of this is that the robot reaches its learned position in the same way despite the slight change in the geometry of the tool (length of the welding wire).

If these changes are not taken into account, the process may be performed incorrectly, which will reduce the quality of the final product or the durability of the tool. In extreme cases, a collision between the tool and the workpiece can occur. A separate tool coordinate system eliminates this problem – it is sufficient to perform a calibration after each wire trimming, which will result in determination of the new, actual end of the tool.

When defining the t_Torch tool coordinate system in this chapter, the “True" parameter was displayed to indicate that the tool is mounted on the robot:

```
PERS tooldata t_Torch:=[True, [[-7,0,352],[0.92388,0,0.382683,0]],
[0.001, [0,0,0],[1,0,0,0], 0,0,0]];
```

You need to know that there are two types of TCP:

- moveable
- stationary TCP, also called Remote TCP, fixed TCP, or external TCP.

Most applications present moveable TCP, i.e. the one that changes its position with the robot’s movement, while the tool is mounted on the robot. Examples of processes with moveable TCP include arc welding, painting or palletizing. All examples used in this chapter concern the moveable TCP.

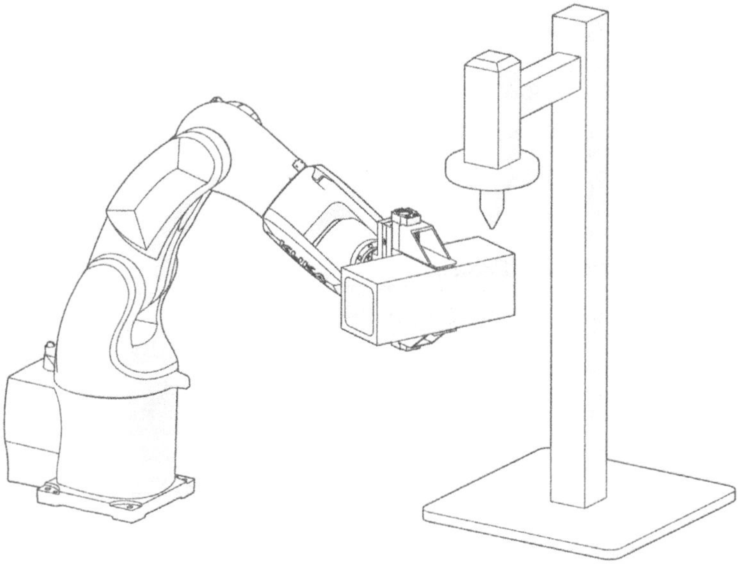

Fig. 54: External tool.

Stationary TCPs do not move when the robot is in motion. It is defined on a fixed tool mounted on the floor around the area of manipulator's operations. Examples of processes that can be carried out with stationary TCP (external tool) include gluing, riveting or spot welding.

The use of a stationary tool permanently installed on the factory floor has its advantages. It is easier to get all the media it needs, and the robot can work with several tools without wasting time on changing them. It is also possible to use a manipulator with a lower payload capacity, because, in most cases, the weight of the workpiece is lower than that of the tool used in the process.

4.4.3. Part coordinate system

A part coordinate system is a reference system that contains information about its distance from the robot's base as shown in Fig. 55. This system is also called the work coordinate system, user coordinate system, user frame or more simply: the base. When this coordinate system is undefined („zero") then coincides with the robot's base system, which you know from chapter 4.4.1.

The part coordinate system is user-defined, and its main purpose is to simplify the programming process. A single robot can have several of those coordinate systems, both for different parts and for the same part in different positions. The use of an additional coordinate system can greatly simplify your work, as you will see in a moment.

Let us return to our example. The robot is still in the same initial position as it was before, but there is a box on the edge of which I defined my own user frame (called *b_Box1*).

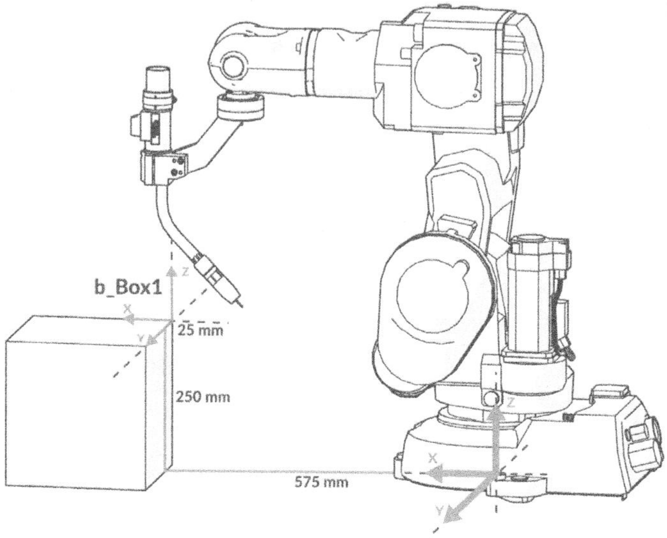

Fig. 55: Part coordinate system.

```
!******** BASE DATA ********

PERS wobjdata b_Box1:=[False,True,"",[[575,-25,250],[1,0,0,0]],
[[0,0,0],[1,0,0,0]]];
```

In ABB robots, the part coordinate system is called *Work Object* and the values are stored in the *wobjdata* constant. Then we have the name *b_Box1*, then come the coordinates X, Y, Z in the square brackets *[575, -25,250]* and the orientation stored as quaternions *[1,0,0,0]*.

If you do not use the user frame, then the robot calculates the position of the point which you want to reach from its base (coordinate system shown in Fig. 45). In the last subchapter our program looked like this:

```
CONST robtarget P20:=[[457,0,295],
[0.382683,0,-0.923879,0],[0,-1,0,6],
[9e+09,9e+09,9e+09,9e+09,9e+09,9e+09]];

MoveJ P20,vmax,fine,t_Torch\Wobj:=wobj0;
```

Now I will add *P30* point, which will stay in the same place (the robot will not move during its execution), but it will use tool coordinate system (*t_Torch*) and part coordinate system (*b_Box1*):

```
CONST robtarget P20:=[[457,0,295],[0.382683,0,-0.923879,0],
[0,-1,0,6],[9e+09,9e+09,9e+09,9e+09,9e+09,9e+09]];

CONST robtarget P30:=[[-118,25,45],[0.382683,0,-0.92388,0],
[-1,-1,-1,6],[9e+09,9e+09,9e+09,9e+09,9e+09,9e+09]];

MoveJ P20,vmax,fine,t_Torch\Wobj:=wobj0;

MoveJ P30,vmax,fine,t_Torch\Wobj:=bBox_1;
```

As you can see, the coordinates of *P30* point have changed significantly, but this does not affect the robot's position (Fig. 56).

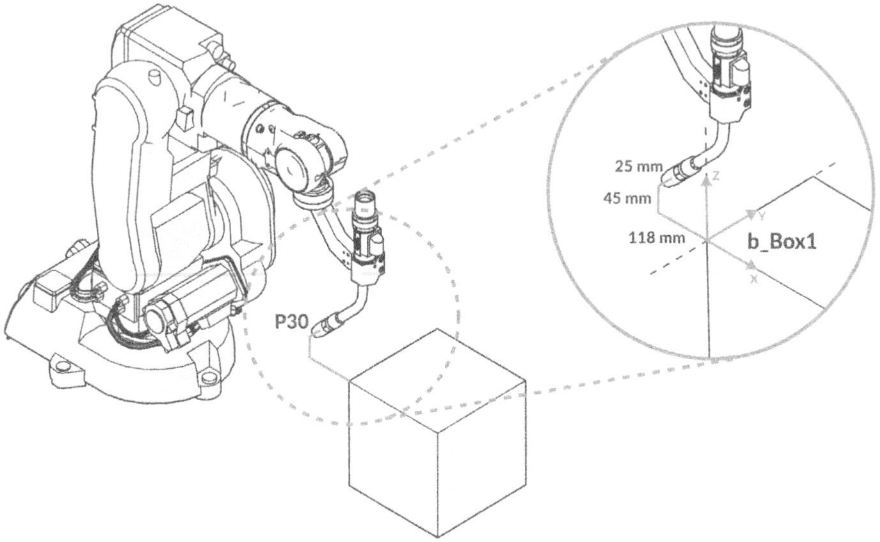

Fig. 56: The coordinates of P30 are calculated from the auxiliary coordinate system.

When using the part coordinate system, the "line of reasoning" of the controller looks as follows: move to the TCP of the tool (distanced from the robot mounting flange by *[-7,0,352]*) to a point that is *[-118,25,45]* mm from the part coordinate system, which is shifted from the base of the mechanical unit by *[575,-25,250]* mm. At first glance, this may seem like its complicating your life, but remember that the part coordinate system has one major advantage over the robot's base system (described in chapter 4.4.1) – it is manoeuvrable. You can freely define its position in space. The distance of points defined in the program (*P30*) from the part coordinate system frame is constant, but the reference system itself can be moved to any place – by entering its current position into the controller.

What are the benefits of defining a user frame? In the case of a robot's manoeuvrability, the advantage is the same as in the case of tool coordinate system. We can move in relation to the newly defined coordinate system. Additional, advantages come to light when we have to modify the program. Suppose that our box has been moved. It is sufficient to update the definition of the user frame and the robot gets back to the same point in relation to the box, we do not need to change the program.

In Fig. 57, I have pictured a situation in which the robot reaches the edge of both boxes respectively (points *P40* and *P50*). Reference frame is defined at the corner of each of them:

```
PERS wobjdata b_Box1:=[False,True,"",
[[575,-25,250],[1,0,0,0]],[[0,0,0],[1,0,0,0]]];
PERS wobjdata b_Box2:=[False,True,"",
[[875,-325,250],[1,0,0,0]],[[0,0,0],[1,0,0,0]]];

CONST robtarget P40:=[[0,0,0],[0,0,1,0],[-1,0,-1,6],[9e+09,9e+09,
9e+09,9e+09,9e+09,9e+09]];
CONST robtarget P50:=[[0,0,0],[0,0,1,0],[-1,0,-1,6],[9e+09,9e+09,
9e+09,9e+09,9e+09,9e+09]];
```

```
MoveJ P40,vmax,fine,t_Torch\Wobj:=b_Box1;
MoveJ P50,vmax,fine,t_Torch\Wobj:=b_Box2;
```

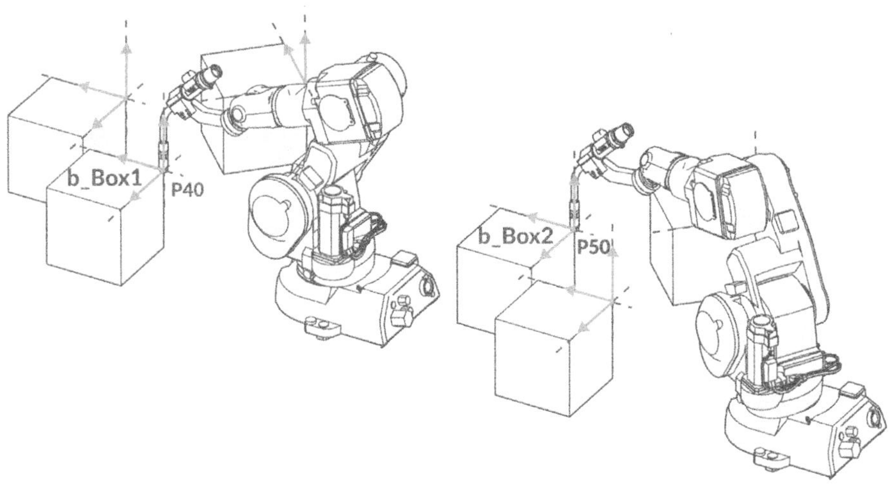

Fig. 57: Advantages of applying the user frame – easy adaptation of the program in case of changes.

As it can be seen from the above diagram, both points (*P40* and *P50*) have the same coordinates, because the distance of the welding torch nozzle from the defined user frame *b_Box1* and *b_Box2* is the same for each case – *[0,0,0]*.

You could try to write a program that would optimise the results that would reach three boxes only by using the one point in the program and three part coordinate systems:

```
PERS wobjdata b_Box1:=[False,True,"",[[575,-25,250],
[1,0,0,0]],[[0,0,0],[1,0,0,0]]];

PERS wobjdata b_Box2:=[False,True,"",[[875,-325,250],
[1,0,0,0]],[[0,0,0],[1,0,0,0]]];
```

```
PERS wobjdata b_Box3:=[False,True,"",[[311.09,-570.71,250],
[0.923879,0,0,-0.382684]],[[0,0,0],[1,0,0,0]]];

CONST robtarget P40:=[[0,0,0],[0,0,1,0],[-1,0,-1,6],
[9e+09,9e+09,9e+09,9e+09,9e+09,9e+09]];

MoveJ P40,vmax,fine,t_Torch\Wobj:=b_Box1;
MoveJ P40,vmax,fine,t_Torch\Wobj:=b_Box2;
MoveJ P40,vmax,fine,t_Torch\Wobj:=b_Box3;
```

The only thing that changes is the position (see at b_Box1, b_Box2 and b_Box3 values) of the user frame. Contrary to the TCP change, it is not necessary to change the coordinates of all the points here. In the example above, the program consists of a single point, which makes it possible to underestimate all the advantages of the additional coordinate system. When you develop a program that consists of 100 or more points, the ability to easily adapt (move and rotate) by changing only one coordinate system significantly accelerates your work.

Imagine that we have written a program (where we use the user frame) in which 10 welds are made. For some reason, the table on which the workpiece is placed must be moved 10 cm away from the robot. The good news is that the welding points are still within reach of the robot's arm. The bad thing is that the welding program is temporarily out of use. If the program points are defined by using robot base system instead of part coordinate system, we have to manually adjust each point to fix it. There is no way to relocate all program points in easy way. They are always defined in fixed distance from the robot base. Thankfully, if we have a defined part coordinate system, it is enough to just modify its position by 10 cm. This single modification will move all points in the program to the right position.

It is easy to adopt programs which were developed offline to the situation on the line by using the user frame. If, in a factory, the table from which the robot picks a part is mounted at a different distance from it than in the simulation, then it is sufficient to enter the actual offset values and the robot will reach the same position as displayed on the computer. On the production line, the position of station elements is measured with the use of industrial laser trackers. Such measurements tell us, very precisely, by how much the reality differs from the virtual simulation.

4.5. Arm configuration

If we consider the simplest 2D manipulator model, it is easy to notice that most of the points in its work space can be reached in more than one way (Fig. 58). In case of a six-axis robot operating in a 3D space, there are many more various arm configurations.

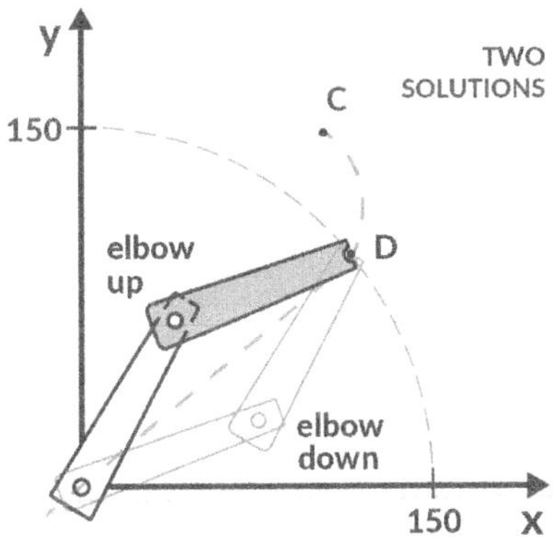

Fig. 58: Different ways to achieve the same point.

A standard six-axis industrial robot can reach most of the points in its workspace in several ways. Although the position and orientation of the TCPs are always the same, the rotation values on the individual axes are completely different. This is a very valuable feature that allows you to reach the target point with different arm configurations, which helps avoid obstacles. However, we need to make sure that the robot reaches a given point in the program every time in the same way, regardless of its starting position. To allow this, In addition to specifying the Cartesian coordinates, it is also necessary to define the configuration of the robot's arms. Syntax example:

PointName (X, Y, Z, Rx, Ry, Rz, *Arms configuration*)

Depending on the starting position and the target position of the movement, some arm configurations are more advantageous than others. It is also often the case, that having several possibilities in theory, may translate to having only one position available in practice – this is due to factors such as external obstacles or the cable arrangement on the robot. The necessity to decide on the right configuration gives the programmer some freedom and allows for creativity, but is, oftentimes, the cause of additional difficulties.

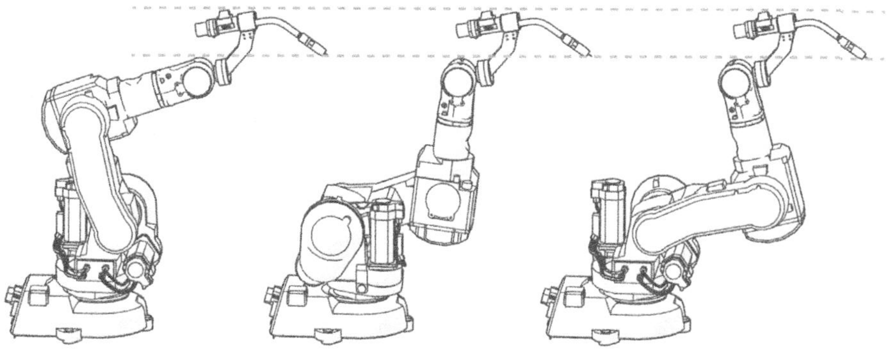

Fig. 59: Despite a fixed position of the tool in space, the configuration of robot's arms is different in each case.

In case of the six-axis manipulator, in addition to a multitude of configurations resulting from a higher number of axes than in case of a 2D robot, one should bear in mind that some joints (such as A4 and A6) have a very wide range of motion, which allows them to compensate each other (adding other possible solutions). When programming, we usually try to keep the robot in the mid-range of the available movements, so as not to expose it to extreme reach positions. This requires more work, but as a result we get smoother movements, shorter cycle times and fewer problems with keeping the manipulator in a good condition.

In the case of a six-axis robot, the configuration is determined with the use of additional parameters. Each robot manufacturer uses its own way of recording this information. The arrangement of the first, third and wrist axes is determined with the help of these records.

Tab. 12: Examples of configuration records used by different manufacturers.

Manufacturer	Record example
ABB	[-1,0,-2,0]
Comau	S (other options: E, W)
Fanuc	N U T 0 0 1
KUKA	S 2,T 35

In Tab. 12, I have collected some examples of different configuration records used by robot manufacturers. Although the record differs for each manufacturer, the rules according to which it is made are the same in each case. Since ABB and Fanuc have the easiest to understand forms of configuration records, we will analyse them.

ABB records the position configuration using the [cf1, cf4,cf6, cfx] syntax, where cf1, cf4, cf6 describe the angular position of axes 1, 4 and 6. The cfx parameter is used to record one of the eight possible arm alignment configurations.

These numbers change depending on which quadrant of the circle the drive is located at. If it is between 0° and 90°, the parameter is equal to 0°; if it is between 90° and 180°, it is 1; if it is between 180° and 270°, it is 2, and so on. If the angle of rotation is negative, the configuration will also be negative. So, a sample record:

```
VAR confdata confg10 := [1, -1, 0, 0]
```

means that:

- the angle of rotation of the first axis shall be between 90° and 180°
- the rotation angle of the fourth axis shall be between 0° and -90°

- the rotation angle of the sixth axis is between 0° and 90°
- the angle of rotation of the fifth axis shall be between 0° and 90°.

In the case of the Fanuc robot, the arm configuration will be saved as NUT 000. The robot operates in 360-degree intervals and, additionally, records the position of the arms with the use of the first three letters:

- N/F - **N**o-flip/ **F**lip indicates whether the fourth axis is in the normal position or rotated "upside down".
- U/D - **U**p/**D**own refers to the position of the third axis.
- T/B - Fron**T**/**B**ack (a little surprise in the naming!) refers to the position of the first axis.

The numbers indicate the rotation value of the drive. The 0 value will appear if the angle of rotation is between -179 and 179 degrees; for the angle range between 180° and 539° it will be 1, and between 540° and 899° it will be 2. If you are in a negative angle range (e.g. -180° to -539°), the value will be negative: -1.

We can carry out several tests to illustrate the effect of different values of the configuration parameter on the arm configuration. We will use Fanuc R-2000iB/165F for testing purposes. We have the following values of rotation angles of individual axes for the NUT 000 configuration:

```
J1: 7.962deg J2: -5.284deg J3: -36.250deg
J4: 77.661deg J5: -51.975deg J6: -79.273deg
```

For the FUT 000 configuration at the same point, the situation will look as follows:

```
J1: 7.962deg J2: -5.284deg J3: -36.250deg
J4: -102.339deg J5: 51.975deg J6: 100.727deg
```

The Cartesian coordinates in both cases look like this:

```
X: 1084.53 Y: 28.68 Z: 181.43 W: 146.63 P: -26,44 R: -21,89
```

As you can see, the sign on the fifth axis has changed, resulting in a change in the rotation angle of axes 4 and 6 to maintain the same orientation. Rotation values of axes 1 to 3 remain unchanged. Note that axis 4 and axis 6 have changed their position by 180 degrees.

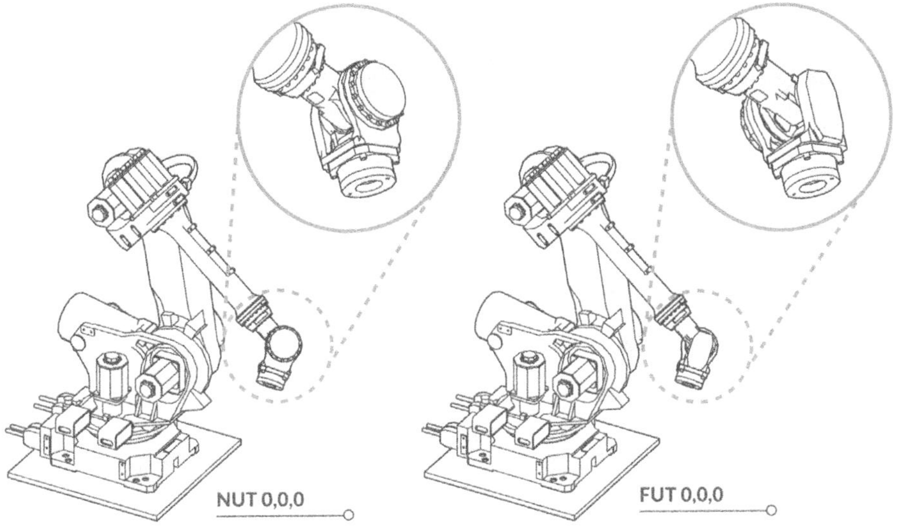

Fig. 60: Change of the arms configuration from no-flip (NUT) to flip (FUT).

At this point, it is worth noting that not every configuration is equally beneficial. In case of FUT, the A4 and A6 axes are closer to the end limits, which means that their rotation values are less favourable. In this particular case, we still have a very wide reserve for manoeuvre before reaching the limit (for this robot, the sixth axis is within the range of -360° to +360°). However, remember that it is worth sticking to the middle of the available range and minimise unnecessary rotations. Moving within the full range available should not be a problem for a mechanical unit, but stretching of the media supply hoses each time in the opposite direction (see chapter 3.4.3), can significantly reduce their lifespan.

Ideally, the robot should be able to run through the entire program in one configuration. If this is not possible (due to existing obstacles or reaching the limit on the axes), reorientation should be planned. Moving to the next quadrant of rotation on axis four or sixth is not a problem, but turning the robot "upside down" on axis three is a complex operation. Any reorientation is a time-consuming process that should be avoided. In case of small spaces, it may be necessary to move the robot away (yet another waste of time!) so that the robot can be reoriented without any collisions.

Pay close attention to the configurations of the subsequent program points. When programming offline, it is very easy to miss some unexpected changes in axis settings (usually 4th and 6th). The same applies to the modification of the robot program in the text editor on a computer. Such an unnoticed change will force the robot to make large sweeping movements, which are likely to result in a collision. It is a very dangerous situation and it should never take place.

Realizing why we record the arms configuration will make you pay closer attention to it. If one of the drives is close to its physical limit, then perhaps it is worth trying to change the configuration at this point. Similarly, if the robot tries to reorient rapidly somewhere in the program then check if the entire program is created using one optimal configuration. Sometimes reorientation is unavoidable, but it should be as smooth as possible. This will save you valuable seconds in the cycle time.

4.6. Singularity

Singularity in robotics is just like an offside in football – it is difficult to immediately get the gist of the problem, because the definition itself does not say much. Formally, the problem could be described in the same way as it was depicted in ISO 10218:

Singularity – condition caused by the collinear alignment of two or more robot axes resulting in unpredictable robot motion and velocities.

Enough with the theory, it is time for practice. There are two problems with singularity. The first one is the inability of the controller to determine an unambiguous configuration of the robot arms. The second is the inability to maintain the linear speed at which the tool moves. For a robot with six degrees of freedom, these problems occur in three arm configurations (roboticsbook.com/singularity/) – the robot is not able to move through these in a linear motion. Such configurations are called singularities and are shown in Fig. 61.

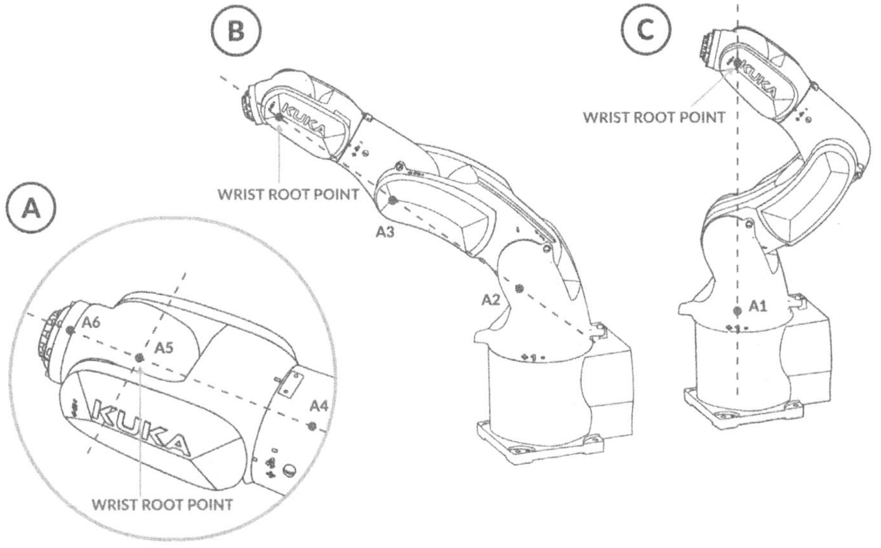

Fig. 61: Examples of singularity configurations for a six-axis robot.

The first, most common case, can be observed when the value of the fifth axis is equal to zero. Then the centres of rotation of the fourth and sixth axes are both positioned in one line. The second case is when the centres of rotation of the second axis, third axis and wrist centre lie in one line. The third occurs when the centre of the robot's wrist tries to move linearly over the centre of rotation of the first axis – this is the least common singularity in practice.

The most common (it can occur practically at any point of the manipulator's working space), and thus the most troublesome, is the singularity of the wrist (Fig. 61A). As you can see, when the fourth and sixth axes are in one line, there is an infinite number of combinations (because in practice we are limited by the angular resolution of the servomotors), which give the same orientation and position of the tool. It can be 0°/0°, 5°/-5°, but also -179°/179° or -105.93°/105.93°. Since both axes (A4 and A6) are in one line, it is possible to level the motion of one servomotor with another one by rotating both of them with the same speed in opposite directions.

This becomes a serious problem when the manipulator tries to follow a trajectory in a linear motion, during the course of which it is necessary to change the sign on the fifth axis from positive to negative and vice versa. When the position on the fifth axis reaches 0°, the robot controller cannot determine the value of the A4 and A6 axes. In addition, since the controller tries to spin A4 and A6 servomotors within the entire range, then in order to maintain the tool's orientation, the speeds they would have to reach are far beyond their physical capabilities. For example, in case of the KUKA robots, the "dead zone" is the range of ±0.0181 degrees on the fifth axis, so we can get very close to zero, but we cannot reach it, neither can we change the sign on the A5 axis during movement.

The position of both axes cannot be unambiguously determined mathematically, by means of an inverse transform. There is any number for A4 and A6 axis generated positions (they are all equally "good"), which sum up to the same angle sum. It is very important to remember that it is not possible to perform a linear motion when the sign of the rotation angle changes on the fifth axis. This is a major inconvenience, especially in applications such as palletising or arc welding. Robot manufacturers are aware of this and offer their singularity-conscious users a set of small "tricks" thanks to which they are able to perform a linear-like motion, that allow for the singularity avoidance.

One of the methods is a slight deviation from the path, which prevents the occurrence of singularities. This can be achieved by treating the section of a linear motion, when the manipulator passes through a singularity area, as a PTP motion. In this way, the robot will not strive to hold the tool orientation unchanged (there will be slight deviations), but it will be possible to change the sign of the angle of rotation of the fifth axis. However, in comparison to fully linear motion, the accuracy of such movement will be lower.

Another method is to limit the rotation range of axes 4 and 6 and reduce the speed of the robot, which allows the robot to move without exceeding the maximum angular speed during the change of orientation.

Another way is to define how axis 4 is meant to behave. Depending on the setting, you can force the robot to keep the value of axis 4 as "0" or hold it in the same position from the beginning to the end of the movement. With one of the variables set to constant, the robot controller must only take care of the correct setting of the sixth axis.

Software solutions allow to "bend" mathematics and physical limitations of the robot's servomotors, but singularities should be primarily dealt with by designing the station and the process in a way that eliminates occurrence of any singularities during the movement. We should not allow the value of fifth axis rotation to be close to 0°. In most cases, we can avoid singularities by changing the height at which the robot is mounted and/or to mount the tool on the robot at a different angle.

Let's move on to a slightly different case. This singularity occurs when the wrist root point (intersection of A4/A5/A6 rotation axes) lies in one line with the centres of rotation of the second and third axes (Fig. 61b). In this case, the robot controller does not have any problem with defining the unambiguous position of its arms, but they are extended to the end of workspace. As a result, even a small linear motion in the Cartesian coordinate system (at low speed) makes it necessary to use very high axial speeds (for the A3 and A2 axes).

It is a good practice to design the workstation and prepare program in such a way that manipulator can move in any direction by 100 mm at any point of his trajectory. This rule concern the medium and high payload capacity robots (smaller units don't have enough reach to obey this principle). If it is respected, then we do not have to worry about this kind of singularity.

The last singularity is similar to the first one and occurs when the intersection of the A4, A5 and A6 axes is directly above the A1 rotation axis (Fig. 61c). The main problem here is that the position of the first axis cannot be determined unambiguously. In addition, there is another problem – this axis needs to be accelerated to a very high speed to maintain the linear motion speed. Since such robot arrangement is quite rare in practice, the problem is less of a nuisance than in the first example.

The singularity problem is related to linear motion (during which the robot tries to keep the tool's orientation unchanged), and it does not occur in the case of PTP movement which can pass through a singularity configuration.

4.7. Types of motion

We already know how we can determine the next robot positions. We know what inverse kinematics, arm configuration and singularity are. With this knowledge we can start thinking about putting our robot in motion. The robot can execute its movement in a variety of ways that have a significant impact on the shape of the trajectory and the cycle time. A movement from A to B consists of: acceleration to the set speed, moving at a set speed, and braking. Industrial robots move by using three types of motion: point-to-point (PTP for short, also known as: joint interpolation), linear or circular.

4.7.1. Point to point motion

It is a basic type of motion in robotics. It is used for all kinds of general movements, where TCP is supposed to get from one point to another, but the shape of the path (e.g. straightness) is not very important. The robot controller tries to make the time needed to move from point A to point B as short as possible, which does not mean that the robot moves along the shortest straight line between the consecutive points of the program. The situation is analogous to that of car navigation, where by choosing the option of calculating the fastest route, the device will try to direct us as long as possible on motorways and expressways, even if it will result in adding a few additional kilometres to the route.

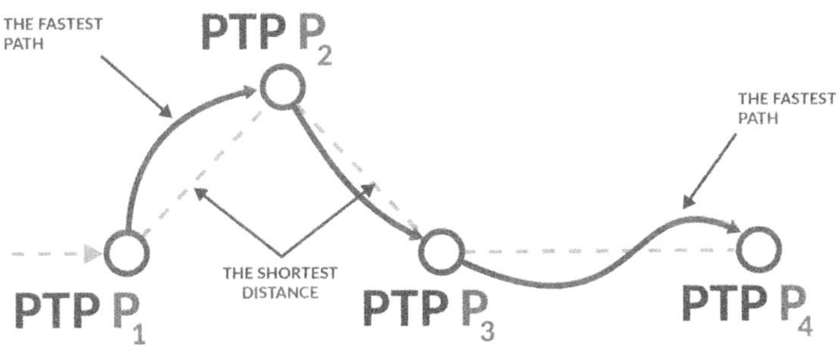

Fig. 62: PTP motion.

The robot works in a similar manner when planning PTP motion. It is more convenient for the robot to give up on the linearity of movement and make a detour, thanks to moving in arcs (just to remind you: all the axes of the robot move by means of a rotation) it can perform the movement faster.

It is not possible to determine the course of the trajectory basing solely on the location points. We will get the information on how the robot's movement will look like only after having it moved along the segment that we want to check. The subsequent segments (provided that the same parameters will be used as during the first time) will be executed on an identical trajectory.

The advantage of PTP motion is its small complexity and lack of problems with moving through singularity configurations. This is due to the fact that the robot does not try to maintain the orientation of the tool – since it changes during the movement. The robot controller calculates the positions of the drives at the final point of motion, and by knowing their initial position, it can determine the angle of rotation of each servomotor.

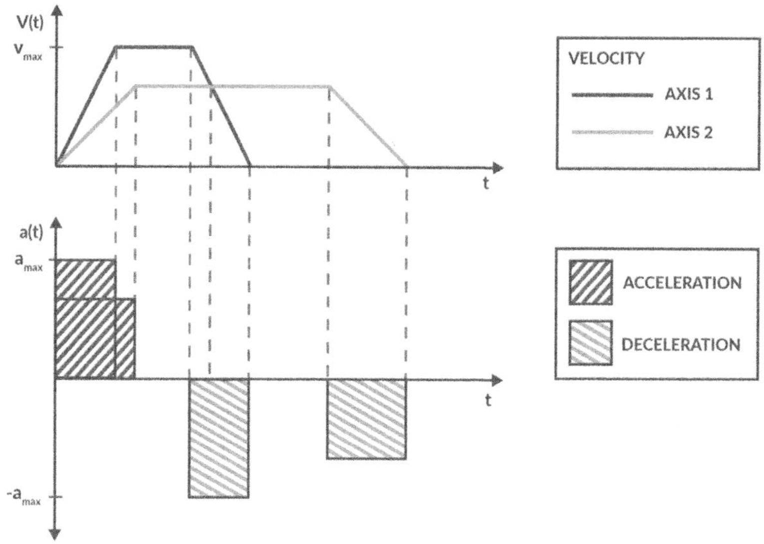

Fig. 63: Asynchronous PTP motion (SLEW).

From a theoretical point of view, two variations of movement are possible from point to point:

- Asynchronous PTP (SLEW)
- Synchronous PTP.

In the first case (*SLEW*), all drives that require rotation start operating at the same time. Each drive moves at its maximum speed without considering the other axes. The drives end their movement at different times, depending on the maximum speed they can achieve and the path they have to move along. This type of control **does not apply to robotics**. The reason for this is quite simple: it is too unpredictable and the movement from point to point requires quite a lot of space.

However, there is one advantage of *SLEW* motion: in theory, using it may reduce the electric energy consumption of the robot. Brakes could be applied on the fastest axis of the robot, allowing it to wait for other drives with the brakes being active, which would have a positive effect on the electric power consumption of the robot. However, this is only an academic theory, in practice the execution of such a task in case of a six-axis robot would be too inconvenient.

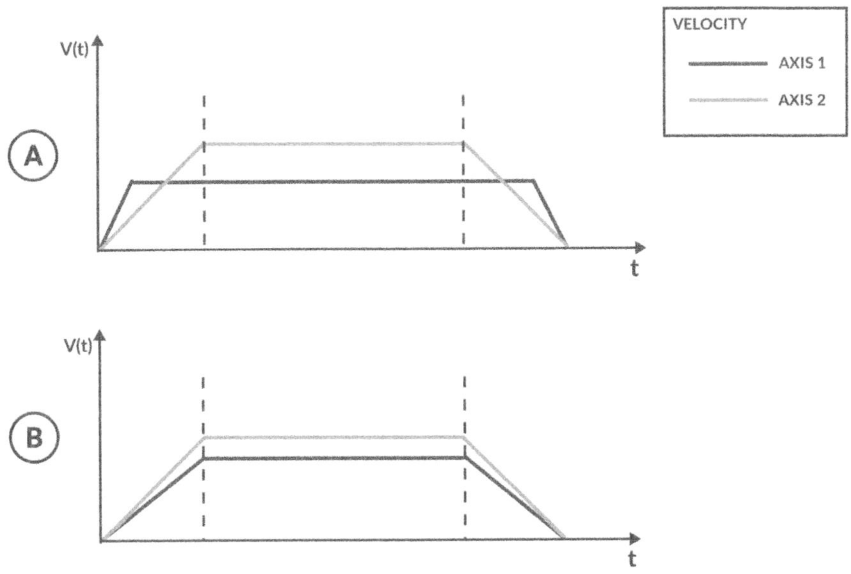

Fig. 64: Synchronised PTP motion.

In synchronous PTP, all drives involved in motion adapt to the slowest axis (the one that needs the most time to reach the target point) so that the motion duration stays the same for all drives (Fig. 64A). The drive that has to move the furthest is called the leading joint and the other servomotors adapt to it.

It is also possible to synchronise the acceleration and braking time, which makes the movement fully synchronised (Fig. 64b). The advantage of synchronous PTP is the smooth, unconstrained and predictable robot motion.

In ABB robots, the *MoveJ* or *MoveAbsJ* command is responsible for the execution of PTP motion, and *PTP* in case of KUKA robots.

4.7.2. Linear Motion

The processes which require a high degree of precision and accuracy (cutting, arc welding, handling of parts) use linear motion. In this mode, the robot controller controls the drives so that the TCP of the tool moves between consecutive points in a straight line and at the same time controls the tool orientation.

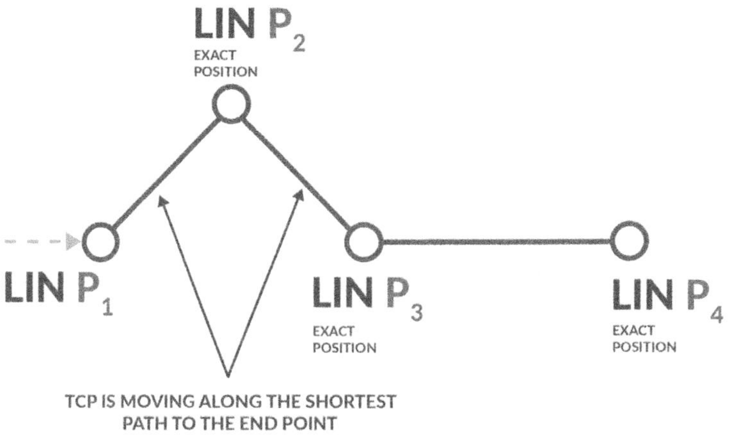

Fig. 65: Linear motion.

This movement is slower than point-to-point (PTP) motion, as there are usually more robot axes involved and not all of them can work at full speed. The faster axes must wait for the slower ones so that the movement can be performed in a straight line. Please note that it is not possible to pass through singularity configurations in linear motion (unless we use the special functions I mentioned when discussing singularities in chapter 4.6).

LIN type of motion is used in narrow spaces or in tasks that require high precision: welding, gluing, applying seals. In linear motion it is possible to control the TCP movement speed which is expressed in distance over time (usually mm/s).

In ABB robots, the *MoveL* command is responsible for the execution of linear motion, in KUKA robots it is the *LIN* command.

4.7.3. Circular Motion

In order to unambiguously define an arc or a circle in space, you need at least three different points which do not lie in a straight line.

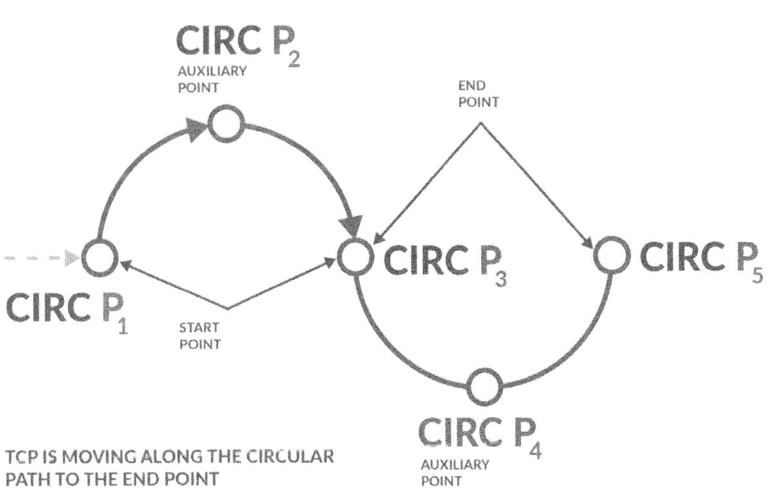

Fig. 66: CIRC Interpolation.

Special attention should be paid to creating small arcs. In such cases, the orientation of the tool often changes significantly during movement. For this reason, it is advisable to move through small arcs at a low speed so that you can react immediately and avoid damage to the equipment. If you are not focused on creating a perfect circle and you do not want to use circular interpolation, you can try to use short segments of linear motion. In the case of arcs with large angles (above 180°), it is a good idea to divide the movement into two or more arcs with a smaller angle.

In ABB robots, the *MoveC* command is responsible for executing a circular movement, in KUKA robots it is *CIRC*.

4.7.4. Motion rounding

The next thing I would like to discuss is the accuracy of the robot movement. Although modern robots have a very high accuracy and precision of movement, it is very rare for them to reach intermediate points precisely. In most cases, they are omitted with a small rounding. It is possible to save a lot of time due to the fact that the robot does not stop at intermediate points. In addition, since the robot controller has a bit of freedom when it comes to the accuracy of movement, from one point to the other, it can make the entire movement much smoother. This has a very positive effect on the cycle time and it makes the robot movement steady.

By forcing the robot manually to drive the subsequent points of the program, we can make sure that the robot will reach them with full accuracy. Rounding can only be visible when you run the program in a continuous mode. Then the command interpreter usually analyses three or four points of the program in advance and optimises the trajectory based on their position.

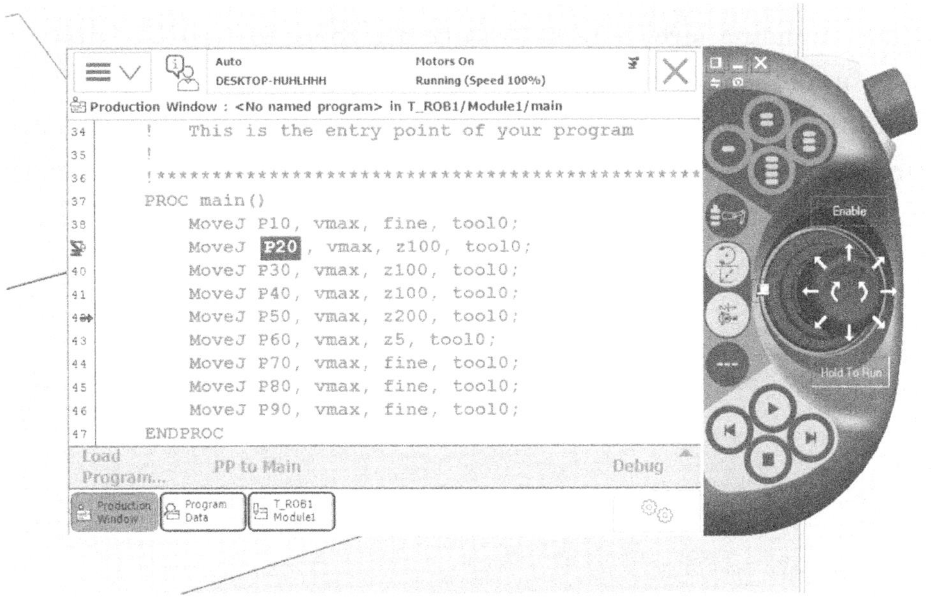

Fig. 67: Example of robot command interpreter's operation.

You can easily see this in Fig. 67, where the robot is in motion from point P10 to P20. The P20 point, which the robot tries to reach (line 39), has been indicated with robot symbol. At this time, the command interpreter is already on line 42 (arrow symbol). The information about the location of P30, P40 and P50 are be taken into account during the planning of the path to reach P20. When you start the program depicted in Fig. 67 on a real robot, you will not notice any difference in the way of reaching the P70 to P90 points, because all of them have accuracy defined as "fine" (i.e. get exactly to this point and stop there). If there were other parameters (like in points P20 to P60: z5, z100, or z200), the robot would behave as shown in Fig. 68.

The maximum rounding is used when the robot moves far away from the parts that it could collide with. When it gets tighter, we reduce the value of rounding, but we try to keep it at high values. Motion rounding is also beneficial in terms of lifespan of a robot.

Smoother movements also put less force on the drives and transmissions than rapid acceleration and braking. In order to reach an exact point, a “fine" rounding level is used to force the robot to stop. The first and last point of the program should be defined in this way. One should also ensure that all other positions, where it is necessary to reach the programmed position (e.g. making welds or picking a part from the table), are also defined in the aforementioned way.

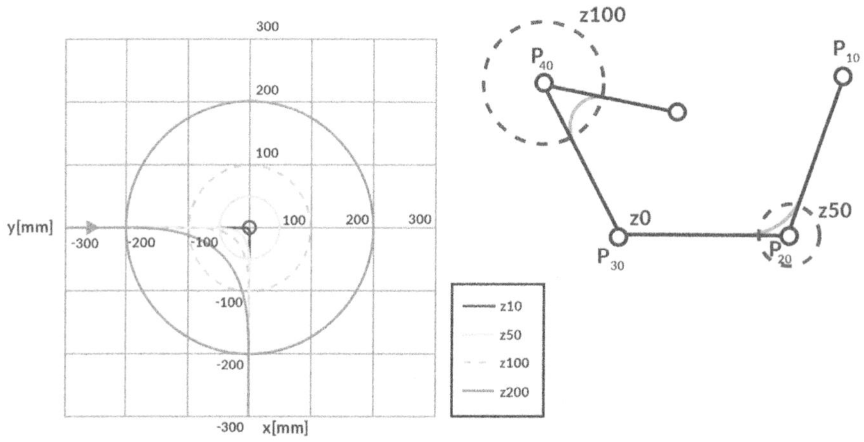

Fig. 68: Example of rounding movement.

Every manufacturer uses a different way of graphing the rounding radius. You must remember that the trajectory on which the robot will move depends on its speed. The fact that the robot avoids collision by millimetres when moving at 20% of the full speed does not mean that a movement executed at a full speed will give the same collision-free result. That is the reason why you should test the programs starting from low speed, and then gradually increase it to allow yourself to see whether or not anything alarming happens.

5

METHODS OF PROGRAMMING ROBOTS

Since the invention of the first robot, there have been many different ideas on how to teach them the tasks they are supposed to perform. The methods used in the initial period were limited by primitive control systems. The first controllers were only able to record the subsequent robot arm configurations. The operator moved the arm of the manipulator to the target positions, by creating a sequence of motion, which was recorded on a magnetic storage. The recorded trajectory was then rendered by a robot.

It was not possible to make any changes to part of the program, or simply run it on any other robot back then. Introducing any kind of modifications to the program meant deleting the old program and starting everything from the scratch. The emergence of the operator panels allowed to control the robot's movement with the use of buttons. In addition, special instructions (controlling the opening/closing of the gripper, handling of external signals) were introduced, which could be put in between the movement commands. The need to robotise complex assembly tasks has ultimately led to the development of more efficient programming tools. That is why the first text-based programming languages were developed in the 1970s.

They gave rise to many new unprecedented possibilities in robotics, which were not attainable before. The programs could be easily modified and transferred between robots. Nothing stood in the way of "borrowing" a part of the program from one robot and using it on another. It was possible to properly document the code with the use of comments. Some of the programming work could be done offline by modifying a text file on the computer while the robot continued to do its job. Systematised data types and consistent syntax have made it much easier to analyse programs. Then the subprograms and functions appeared, which speeded up the programming process.

Over time, technical development has allowed for even more complex human-machine interactions, which have had a significant impact on the popularity of different programming methods. Nowadays, robots are programmed using one of these three methods:

- offline
- online
- hand-guided

Each of these methods can be used to write a program in a form that can be interpreted by the robot. Robot programs usually come in the text form and their syntax is quite similar to computer programming languages such as Pascal, Basic or C.

Below you can see a part of a program written in RAPID language for ABB robots, which is responsible for the robot's movement to the picking position on the table "ST050" where the part is located.

```
MoveAbsJ HOME,vmax,fine,t_Grp01\WObj:=ST050;
MoveJ LHP005,vmax,fine,t_Grp01\WObj:=ST050;

!Pick position
MoveL Pick_st50,v200,fine,t_Grp01\Wobj:=ST050;

GrpPartChk 1, GR01_B01PP\Part2:=GR01_B02PP\Part3:=GR01_B03PP\
Part4:=GR01_B04PP;
GrpSet Cls, Grp_11\Grp2:=Grp_12\Grp3:=Grp_13;
GripLoad_CTRL\Normal_Run:=ld_PT_F57\Dry_Run:=load0;
PLC_Release "Request OK to Pick Part"\Set_:=PLC_do_User_01\
Wait_:=PLC_di_User_01;
SetRobotTypeCode\GripperSystem:=1\PartOnGripper:=TRUE\InPos;

MoveL LHP010,v200,fine,t_Grp01\Wobj:=ST050;
MoveJ LHP015,v200,z200,t_Grp01\Wobj:=ST050;
```

For comparison, below is a comparable program written in KRL language used by KUKA robots.

```
PTP HOME Vel=100 % DEFAULT
PTP LHP005 Vel=100 % PDAT2 Tool[10]:GRP_01_O Base[1]:ST050
; Pick position
LIN Pick_st50 Vel=0.2 m/s CPDAT1 Tool[10]:GRP_01_O Base[1]:ST050

WAIT FOR ( IN 19 'Part 1 present' ) EXOR ( Ghost () )
WAIT FOR ( IN 20 'Part 1 present' ) EXOR ( Ghost () )

SK1_CLAMP_QM1_CLOSE ( )
SK2_CLAMP_QM2_CLOSE ( )

OUT 3322 'In Position Pick ST050' State=TRUE
OUT 3377 'A3377:Part 1 in gripper' State= TRUE
WAIT FOR ( IN 3322 'E3322:Request OK to Pick Part' )
OUT 3322 'A3322:In Position Pick' State= FALSE

LIN LHP010 Vel=0.2 m/s CPDAT9 Tool[11]:GRP_01_M Base[1]:ST050
PTP LHP015 CONT Vel=100 % PDAT10 Tool[11]:GRP_01_M Base[1]:ST050
```

Despite the syntax differences, both programs are quite similar. You can easily distinguish movement instructions (*MoveJ* and *PTP*), the names of subsequent motion points (LHP005 ... 015) and the parameters of this movement. These include rounding (*z200* and *CONT*) and velocity (*v200* and *Vel=0.2*). There are different ways for tool and base recording, and there is also a different punctuation mark that indicates the start of a comment (ABB – exclamation mark, KUKA – semicolon). If you know the syntax of any robot programming language, it is quite easy to for you to decipher the syntax of other languages, but creating programs that are more complex than moving from one point to the other requires a deeper understanding all the commands and data types available.

Tab. 13: Syntax comparison for different robot programming languages.

Robot programming language	Joint Move	Linear Move	Speed Set	Path Rounding
ABB Rapid	MoveJ	MoveL	speeddata	zonedata
KUKA KRL	PTP	LIN	VEL	APO_DIST
Fanuc KAREL	J	L	speed	CNT
Yaskawa INFORM	MOVEJ	MOVEL	VJ/V	PL

Timing of execution of the subsequent commands by robot is strictly related to its position in space. In the example above, the gripper will only close (*GrpSet Cls* instruction) when the **Pick_st50** position is reached. There are instructions that can be executed without considering the robot's current position, but in most cases we want a particular event to take place, whenever the robot is in a given position, that has been programmed in the workspace in advance. Please remember to stop the robot at a point where you want it to carry out a task (e.g. pick parts) and wait until the task is completed (e.g. closing of a gripper). Otherwise, there is a high possibility that it can result in a collision.

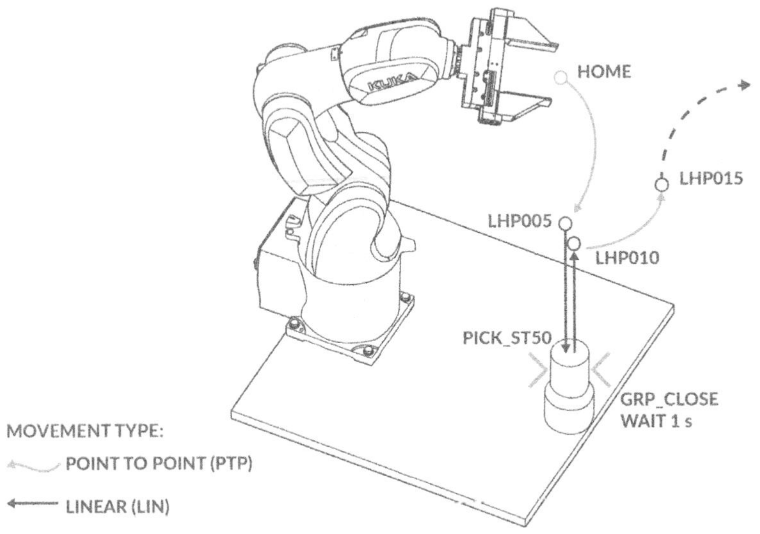

Fig. 69: Program sequence.

Of course, modern robots can easily handle multitasking and they can additional tasks in the background. However, the robot's work characteristics make it execute the program sequentially. It is worth remembering that the robot command interpreter runs a few instructions forward and analyses their contents (chapter 4.7.4). This is necessary for rounding of a movement.

In "classic" programming (writing applications for PCs) and in robot programming, the comments are of great importance. They should be concise, legible and describe very well a part of a program that they relate to. Oftentimes, if a program has not been finished/tested to the end, a logical condition check is added at the beginning of the program:

- Wait for False (KUKA)
- Wait until FALSE; (ABB).

It serves as a protection against starting the program in automatic mode. If the operator wants the robot to make further movement, his or her intervention is required.

If such command appears in the first line of the program, one should write why it is there, where it came from, and what still needs to be improved. This will make it easier for another person to continue our work. All relevant process points should be preceded by a comment (it could be their name, process parameters, etc.). It is a good practice to write comments in a language that can be understood by most people, which, in the case of the automotive industry, usually in English or German.

5.1. Online programming

This method of programming is performed on the production line – the programmer stays in close proximity to the robot and creates a point by point program with the use of a control panel.

By moving the robot and saving the consecutive points of its trajectory and adding appropriate conditions and logical instructions, we can create the program ourselves. The online programming time can be shortened if you don't have to create a program from scratch. We can use previously prepared programs: generated offline with appropriate software, upload copies of a similar program from another robot or process an existing program (e.g. by making its mirror).

Whatever we choose to do, the result will be the same – we get a robot program that requires testing. In theory, the program written by using teach pendant, which requires the least amount of work. It has all the necessary instructions needed and its trajectory has been verified in terms of a collision-free movement from point to point. The next thing to do is to test the program on the subsequent loops at a higher speeds, which will allow you to verify that there are no errors in the logical instructions, and check how the robot rounds the trajectory while moving in a continuous motion. It may turn out that the path will need to be corrected due to the rounding features performed by the robot's movement.

We follow the same steps when we have an uploaded backup or a program generated on the basis of an existing one. When testing programs that was developed offline, it is usually necessary to update them with the instructions missing and verify whether the logical part is correct. It is also necessary to move through the program carefully, starting with point to point movement, and then in a continuous mode and at increasing speeds.

As you can see, online programming allows you to verify the programs that you have obtained in any way. Moving through the program in manual mode and paying attention to the robot's behaviour ensures that there will be no collisions with the station equipment anywhere on the way, and that the logic of the program is correct as well. We are equipped with all the elements necessary to achieve this goal. We see the robot in its working environment, together with all the equipment that the robot may collide with during the process. We hold the control panel in our hands and we can see the entire program of robot's motion along with all the parameters, which can be modified at any time so that we get an optimal result. Teach pendant displays information on all errors and warnings.

Online programming gives you the access to all robot features. However, the verification of the program in real-life conditions is time-consuming and requires the production to be halted. It is also not possible to prepare the program in advance since all the changes must be introduced after the robot has been installed at the workplace. This method requires the presence of the programmer at the factory, which makes it impossible to program the robot remotely.

The advantage of online programming is the fact that the resulting program is 100% correct and it enables a high-quality process execution. It allows you to prepare programs with the consideration of the quality dispersion and measure the actual cycle time, by taking into account the actual opening and closing times of the grippers, time losses associated with signal exchange and all stoppages that occur during the production.

5.2. Offline programming

The need to avoid long production line downtime in the case of online programming, has resulted in the creation of an offline programming method. The program is generated on the computer without the need to have the real robot in proximity. The most basic form of robot programming is using a text editor and then modifying the program files manually. This is a very wearisome job during which the programmer must focus and predict the robot's behaviour. From a practical point of view, it is virtually impossible to create a program from the scratch in this way. Nobody is able to calculate the coordinates in 3D space in a way that would allow them to visualise the position of the robot solely on their basis. With a text editor you can successfully add logical instructions, modify parts of the existing programs (mirroring, renaming points, changing tool or part coordinate systems definition). There are special text editors for robot programmers such as OrangeEdit for KUKA robots or a built-in editor in ABB RobotStudio which make the work easier.

Nevertheless, the simulation programs have numerous additional options. The ability to visualise a robot in a 3D environment allows you to significantly improve the offline programming process. However, the computer screen is not the only way you can see the future robotic station. You can also take a walk on a digital model with the use of VR headsets (roboticsbook.com/VR/). By using a simulation software we have more opportunities to optimise the program and detect errors at an early stage. It is possible to design the product and the production line simultaneously. Design errors can be identified quicker, even before any equipment is built. These are the advantages of offline programming over the online one:

- program preparation may take place without the need to stop production
- increase of the programmer's work safety, since the time spent in robot's proximity is significantly reduced
- visualisation of the robot's work in a 3D environment helps identify errors at an early stage of the program preparation
- production optimisation is made possible with an easy testing of the various scenarios
- possibility of reusing old programs (full or parts)
- collision-checks
- analysis of the process point reachability
- verification of the robot configuration and of the potential problems with arm arrangement (axes limits, singularity).

Nowadays, offline programming is the final stage of a bigger process. At present, it is very rare for a robotised station to be rough built. Instead, a specialised software is used, which allows you to build a virtual model and make sure that everything will work in reality, which saves you a lot of time and money.

Fig. 70: ABB RobotStudio program.

We can distinguish three phases of robotic station construction: pre-construction phase, simulation and offline programming. At the first stage, general assumptions are made: the station concept is developed, taking into account the limitations of the process, the space limits in the production hall and the number of robots needed. When we figure out what we need and in what quantities, we can move on to the simulation phase. During the simulation, the designers develop table and gripper concepts. The robotics engineers check these concepts on an ongoing basis and select the remaining tools needed to perform the process. The relative position of the robotic station elements is determined. If there are any problems, the engineers try to solve them together. If it is possible, the designers will introduce changes to the design, the robotics engineers may decide to replace a tool used in the process or redistribute process points. Then the correctness of the assumptions made in the pre-construction phase is verified.

The final stage of the simulation is when all process points in a virtual environment could be accessed without collisions, and the existing preliminary paths allow for the planned cycle time not be exceeded.

The last stage is OLP (*Offline Programming*), in which the robot's motion paths get their final look. The signals for communication with the PLC start appearing, the customer naming standard is introduced, the information about the collision zones is entered, and the information regarding the tool and base coordinate systems is inserted.

One disadvantage of offline programming is that it represents the world in an idealistic way. In simulation programs, the workpieces are always perfectly geometric, the sheets are not oily and there is no gravity. Dynamic parameters such as tool vibrations, robot arm deflections, the influence of load on trajectory quality are not taken into account. Many processes cannot be reproduced correctly on the computer screen. It is not possible to set all parameters (e.g. for arc welding) or to display the current preview (roller hemming). In this case, it is possible to check the movement reachability and estimate the time of the movement, but the final path will still have to be, for the most part, set by a robotic engineer working online. The programs generated offline needs to be verified online in all cases.

In the absence of a current update of changes taking place in the real station in the factory, very quickly programs created offline become useless. Any mechanical modification to the equipment should be reflected in the digital factory model. In order to improve this process laser measurements and 3D scanning are used.

5.2.1. Simulation programs

When it comes to programming robots, time is the most precious resource. Developing programs, verifying them, connecting them with the other peripherals and testing takes a lot of time. Time which could have been devoted to production. Of course, you can integrate new robot programs at weekends or overnight, but this increases the costs and the number of human errors. The scope of work is sometimes so comprehensive that just the preparation time alone requires a few days of downtime on the production line. That is when the offline programming, supported by the appropriate applications, comes in handy.

The simulation programs that are available on the market differ significantly in terms of its functionality. Applications provided by robot manufacturers support only their own products, but they are giving access to all features built-in into the robots.

Tab. 14: Simulation programs offered by robot manufacturers.

Manufacturer	Program title
KUKA	Kuka.Sim
ABB	ABB Robot Studio
Fanuc	Fanuc Roboguide
Comau	SMART Sim 3D-Lite
Kawasaki	K-Roset
Yaskawa	MotoSim
Stäubli	Stäubli Robotics Suite
Panasonic	DTPS-G3
Nachi	AX On Desk

The second group consists of programs that allow for simulation of robots of different brands in one virtual station. Due to the fact that it is possible to design workstations equipped with robots from different suppliers, it is enough just to be familiar with one application interface to program them all. RobotExpert, Delmia, Process Simulate and Robcad are the most popular applications of this type. Their disadvantage is that the support for less important functions available on robots is limited, and not all devices are always available in program libraries. RCS modules are used for making very accurate simulations.

In the past, these programs were classified under the CAR term (*Computer Aided Robotics*), which was coined on the basis of CAD acronym (*Computer Aided Design*) or CAM (*Computer Aided Manufacturing*), however, this term did not catch on.

5.2.2. RCS/RRS

The beginnings of robotic simulation software were difficult, and the low performance of computers back in 1980s was not the only reason for this. The first programs that were introduced to the market were not related to any robot manufacturers. Whilst it was possible to check the reachability and verify the occurrence of static collisions, any attempts to measure time or plan trajectories accurately were far from ideal. In order to calculate the robot's motion, general control algorithms were used and the results they produced were very different from those of actual controllers. Deviations in time and trajectory when considering a real robot and a simulation amounted even up to 30%.

Currently, every major robot manufacturer offers a simulation software with built-in algorithms that control the robot movement. It is not a problem, since no information is passed outside of the company. People writing a simulation and offline programming software sits two desks away from those who created algorithms for the physical controller. However, this turned out to be a problem when the independent software manufacturers (such as Siemens Industry Software, Dassault Systèmes) asked for this information. The robot manufacturers feared that the information which is regarded as business secrets will be revealed.

Due to a lack of information, the developers of tools such as Robcad tried to obtain all data they needed by studying the behaviour of real robots. The lack of information such as braking and acceleration times, inertia of servomotors, the behaviour of the manipulator carrying a load did not allow to accurately reproduce the actual robot movement. For this reason motion planners were written by using reverse engineering. It was an arduous work, which has not resulted in the achievement of satisfactory results. This method was far from ideal from the beginning, but the situation was becoming even more complicated with every new generation of controllers. Let's take a look at the comments that are included in Robcad 3D robots models:

```
*
*Robot : ABB IRB6000/2.4-120 and IRB6000/2.4-100
*Controller: S3 - Versions M92 and M93
*Controller ID :
*Date created: April 1993
*Created by: Udi Naamani, Noam Ribon and Mike Hatwell
*Status: Incomplete
*Level : Documented
*Updates :
*- Originally written by Sam Yechiam (January 1992) and based on
*measurements made at ABB Detroit, USA.
*Dimensions taken from "Product Manual - IRB6000" (unverified)
*Max speeds " " " """
*Limits " " " """ "
*- Updated by David Aufhauser (June 1992)
*- Updated by Udi Naamani (September 1992)
```

```
*- Updated by Ray Kliemann (October 1992) with measured
cycle times
*using a 100 kg load
*- Updated by Ray Kliemann (March 1993) with jerk parameters
*- Updated by Ray Kliemann, Noam Ribon and Udi Naamani
(March-April)
*1993) to include:
*(a) extensive measurements made at ABB USA and CTC Robotics Lab.
*(b) new motion planning software written by Mike Hatwell
*- Updated by Mike Hatwell (October 1993)
*- Updated by Mike Hatwell (April 1994) to adjust the location of
*the measurement matrix relative to the BASEFRAME
% begin text

TECNOMATIX has performed extensive testing to verify the accuracy
of this model. Nevertheless, the results are only experimental and
by their very nature, dependent on specific tasks and measurements.

% end text
```

As you can see, this digital model was created in stages, the geometric dimensions were taken from the documentation, which may have been out-of-date. The tests were also carried out partially; when a new version of the simulation algorithm appeared on the market, new measurements would have to be taken separately in order to get a proper reflection of the reality.

There were attempts to develop universal robot control algorithms, but the progressing specialisation and complexity of real algorithms has always stood in the way. All this has finally forced the robotics industry to start working on a common standard. Audi, BMW, Ford, Mercedes-Benz, Opel, Peugeot, Renault, Volvo and VW, along with software and robot manufacturers, started the "Realistic Robot Simulation" (RRS) project in January 1992. The first version of the standard first came to light in January 1994 and it allowed to achieve a very accurate representation of the motion and time of virtual robots.

The solution that was invented works in a relatively simple way. Robot manufacturers supply RCS (Robot Controller Software) modules in the form of "black boxes" – their content is encrypted. The simulation software sends requests to the RCS with the help of imposed commands and gets answers in the same form. There are less than 170 of these standardised commands. About 100 of those are responsible for handling and synchronisation of errors, and only 70 of them are related to robot motion and its parameters.

Only the developer (*robot manufacturer*) knows what happens to the data between the input to the RCS module and its output. If the developer is the author of a sophisticated algorithm (e.g. realising linear motion and using methods to prevent singularities), there is no fear that a competitive company will use this solution. That is exactly how the simulation software developers have gained access to information on how the robot should move, while at the same time, the manufacturers of these robots had no reason to fear of somebody stealing their algorithms. The RRS interface significantly increased the quality of simulation – the deviation between simulation and reality amounts to 3%. Without the RCS module, this value is approximately 30%. All other important parameters of manipulator movement simulation have also improved: cycle time estimation, trajectory accuracy, acceleration profiles, behaviour in areas close to singularities.

Tab. 15: Selected commands from the RRS standard.

INITIALIZE	MODIFY_RCS_DATA
RESET	SAVE_RCS_DATA
TERMINATE	GET_INVERSE_KINEMATIC
GET_ROBOT_STAMP	GET_FORWARD_KINEMATIC
GET_RCS_DATA	GET_CELL_FRAME

SELECT_FLYBY_MODE	SET_MOTION_TIME
GET_MESSAGE	DEBUG
SET_INITIAL_POSITION	SET_NEXT_TARGET
GET_NEXT_STEP	SELECT_MOTION_TYPE
SET_JOINT_JERK	DEFINE_EVENT

The RRS solved the problem of projecting motion accuracy, but issues such as signal handling remained unresolved. For this reason, in April 1994, the RRS II project (VRC – Virtual Robot Controller) was launched, which resulted in publishing of the second standard version in 1998. It supports signal processing, native syntax of the robot program and emulation of the file system. Its disadvantage, in comparison to RRS, is that it requires more computing power and is more difficult to implement due to greater complexity.

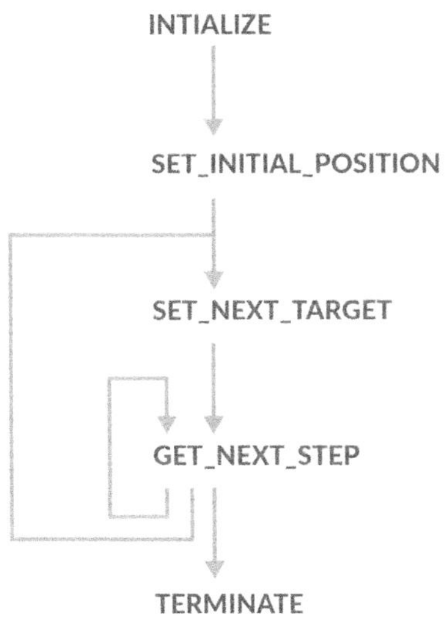

Fig. 71: Scheme of RCS operation.

The newest feature in simulation and offline programming software is virtual commissioning. This software, equipped with RCS modules, gives the possibility of very accurate simulation of individual robots. Nowadays, we want to be able not only to analyse the individual stations, but also the entire production lines, with all the dependencies that can be found there (i.e. mixed production, different product variants, maintenance programs and executing sequences of programs in a loop).

In the real world, the PLC controls all machines on the production line. The PLC sends commands to the robots which indicate which program is to be executed and monitors the work progress. The sequence of tasks programmed in the robot must correspond to that of the PLC.

Modern software used to simulate the robots' work (installed on a PC) allows you to establish a connection between the robot and a physical PLC. Programmable Logic Controller in this scenario can communicate with all virtual devices present in the simulation in the same way it would with their physical counterparts later on the production line. Communication goes both ways, the PLC receives running feedback about the progress of the production process.

This allows to verify whether the code executed by the PLC corresponds to the tasks performed by the robot, which significantly reduces the debug and commissioning time on site. All this takes place long before the physical installation of the equipment on the production line. The benefits of virtual start-up include easy verification of correct signal exchange, verification of the sequence of subsequent production steps, validation of visualisation on control panels, and the possibility of testing of handling emergency situations at an early stage of designing a robotised station.

5.3. Hand-guided robot programming

Hand-guided programming is currently the least popular method among those listed in this chapter. It could be classified as one of the variations of online programming, but instead of using the control panel, learning takes place by grabbing and dragging the robot arm to the desired position and saving its position and configuration.

This method had its golden age in the initial period of robotics development. Machines from that era did not have their own programming languages. The first control systems recorded the program on magnetic tapes, which required rewinding after moving through the program. As a solution for this limitation, a sets of two tapes were used. While the robot was performing a program written on one of them, the other one was rewound to ensure the robot was running in a continuous mode.

In this way, the first painting robot was programmed: Trallfa TR-2000. In addition to direct arm dragging, there were figures of these robot models (phantoms), which with the help of built-in sensors were able to determine their position and pass it on to the real robot. Thanks to that it was possible to move heavy, high-capacity robots that would require a great deal of force from the operator in case of trying to move them manually in order to generate movement. The progress in the construction of control systems was a chance to move away from this inaccurate method.

A few years ago, the programming of robots by hand guiding was practically non-existent in industrial practice. Online programming was more accurate and offline programming was faster. Only recently this method has been gaining popularity due to the emergence of collaborative robots on the market. Hand guiding is currently precise enough and it is also the easiest method of programming to learn for an operator. It does not require technical knowledge and allows the production employees to teach the robot. KUKA LBR IIWA and Universal Robots UR5 are examples of a modern robot that can be programmed in this way.

Nowadays, manufacturers also offer this programming method for larger arc welding models. This allows the robot to be operated by a welder or a machine operator without the need to hire a robot programmer.

In case of Motoman robots, the Kinetiq Teaching solution from Robotiq needs to be used. In order to teach trajectory by dragging the robot arm, a special module must be installed on the robot wrist, which can detect small forces acting on the welding torch and move the whole arm in accordance with the force applied. Solutions based on force and torque sensors mounted on the robot wrist allow you to easily add a new programming method to classic designs. In this way you can move high-payload capacity robots (solutions: KUKA ready2_pilot, Fanuc Hand Guidance, SafeGuiding from MRK or OptoForce Hand Guide) by gently pushing it in the desired direction.

I can imagine that in a few years this programming method will be a standard among robots designed for the purposes of arc welding. Nowadays, many companies struggle with finding welders to work, so it is more and more common for the production in such plants to be robotised. However, automation does not solve all of our problems, because, on one hand, we still need people with welding experience (to control the quality and enter the correct welding parameters), and on the other hand, we need programmers to correct the robot's trajectory. Robot programming by demonstration significantly lowers the entry barrier and allows the welder to quickly acquire the skills of a robot programmer.

The SRP Teach Handle presented by ABB in 2015 is an interesting solution. In short, it is an imitation of a real paint gun, equipped with sensors that allow you to read its position in space. This allows the operators to program robots in a way that they regard as the most natural for them. The worker moves the gun in the same way he would when painting a workpiece with a real tool. No specialist knowledge of robotics and control is required, and it is also easier for the operator to create a smooth and unconstrained path needed to apply coatings without the formation of paint stains. The program saved in this way is passed on to the robot, which reproduces the saved trajectory. In a sense, the history has come full circle, as the first painting robot from Trallfa, which over time became a part of ABB, was taught in a very similar way.

6

MULTI-ROBOT WORKSTATIONS

Robots work less and less independently. It is a standard now that there are stations consisting of many manipulators cooperating with each other. Manufacturers try to produce even more compact manipulators, which allows them to fit a greater number of devices in the same area. Their aim is to reduce the weight of the tools as well, which allows for the use of robots with a lower payload capacity and a slimmer design.

In the automotive industry, the highest density of robots can be found in a framing station, where the underbody and the body sides of a car are connected to one another. The Comau Open RobotGate Solutioion developed by Fiat, can fit up to 18 robots that work simultaneously in a small space (roboticsbook.com/framing/).

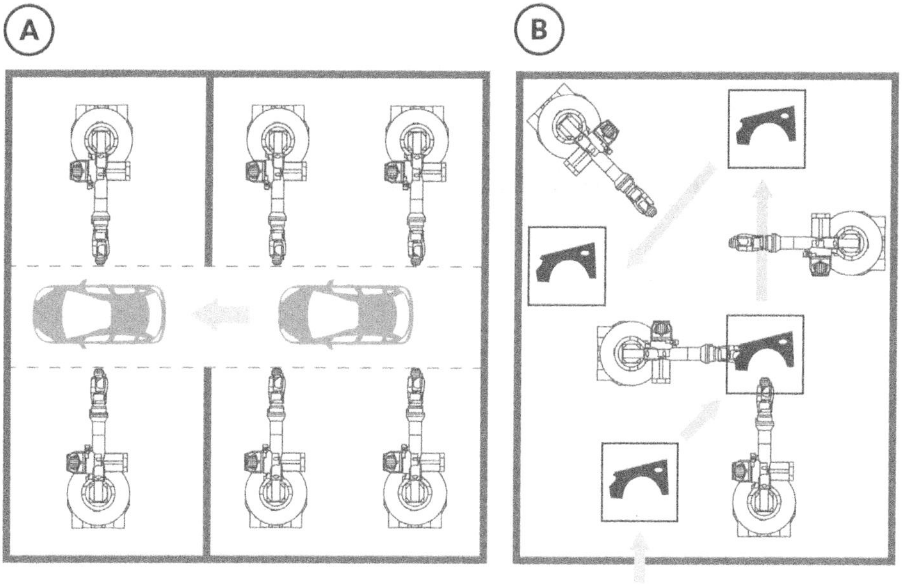

Fig. 72: Robotic assembly: a) in line b) cell production system.

Two types of robotic assemblies can be identified. The first one of those, assembly in a line, is known mainly from the final stages of car production, where the car moves on skid with robots located on its sides (Fig. 72a). There is also a cell production system, in which the robots are in various positions, and the parts are moved from one station to another by robots or they are transported by conveyors (Fig. 72b). Working with multiple robot workstations requires much more attention and accuracy during the station planning and the design phase.

Theoretically, as the number of robots increases, it is possible to achieve shorter cycle times. Splitting tasks between more devices means that each of them has less work to do and can finish it faster. In practice, the problem turns out to be much more complex. In case of poor process planning (separation of process points and sequence of tasks) and small spaces, it may turn out that the robots do not work at most times, because they are waiting for each other until one frees up the work space for the other.

More and more often two robots perform the same task together at the same time. An example of such cooperation can be transferring of the entire car bodies by a tandem of two manipulators or arc welding, where one of the robots acts as a positioner/rotator, and the other, at the same time, welds the element held by the first robot. More and more often one controller is responsible for controlling more than one manipulator and additional rotor axes (solutions: ABB Multimove, KUKA RoboTeam, Fanuc Dual Arm/MultiArm). Some robot models are developed by putting two smaller units together (Yaskawa SDA5F – 15 axes), which are controlled from one controller.

An interesting application is the riveting of aircraft plating, which until recently was done manually. It requires drilling a hole in the plating, inserting a rivet and bucking it. During this process, the rivet is stamped on the other side of the plating. This monotonous and strenuous work requires the coordination of two people working on the outside and inside of the fuselage. The Boeing factory has implemented the *Fuselage Automated Upright Build* process (FAUB for short), where two robots work together on independent platforms on both sides of the plating without the need to see one another. 50,000 rivets are assembled in Boeing 777 in this way. The coordination of two manipulators, one working inside the fuselage and the other one outside, is a huge challenge (roboticsbook.com/Boeing/). On one hand, the collisions with aircraft components must be avoided, and on the other hand, the work must be carried out simultaneously by both manipulators on both sides of the plating.

6.1. Interlocking zones

The basic problem in a multi-robot environment, which the programmer has to face, is the synchronisation of tasks carried out by individual robots. At the very early stage of programming, one should be able to answer the question which robots can work simultaneously and which cannot. If a robot cannot work with another robot synchronously, one should make a decision which robot will go first. What impact will it have on other robots working at this station? What will happen if one of the robots fails and stops during operation? Is there a possibility that one of the other operating manipulators will hit it? Can any robot perform maintenance work while the other robots are in operation during production? One should find answers to all those questions.

In order to prevent collisions, it is necessary to designate all areas of common workspace (interlocking zones) for all manipulators working simultaneously. Fig. 73 presents an example of such zone (hatched area). If robot "A" wants to work in the hatched area, it has to ask the PLC for permission before entering this area (Fig. 73A). If there are no obstacles (the zone is free and no other robot is working there at the time), the robot will get permission to continue its movement. Robot "A" will then continue to execute its program, and, at the same time, it will send the information to the PLC that the collision area is now occupied (Fig. 73b).

The same procedure applies to robot "B". Before entering the collision zone, it needs to ask for an entrance permit. Until the PLC will not receive the information from robot "A" that it has left the common work area, the PLC will not let robot "B" proceed with its movement. Robot "B" will have to wait until robot "A" finishes its work in the collision zone (Fig. 73c).

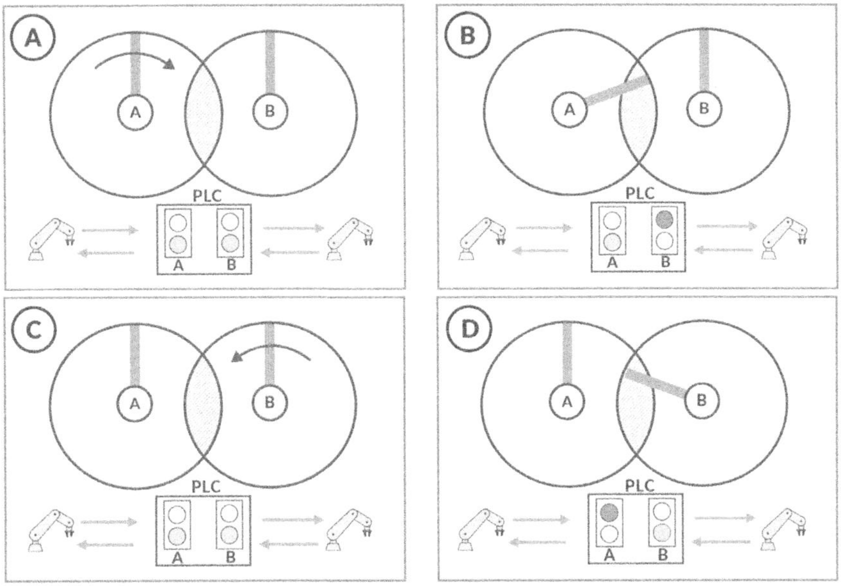

Fig. 73: How the interlocking zone between robot A and B works.

It is worth to remember about designating interlocking zones not only in case of the main program executed by the robot, but also in case of robot maintenance programs (tool change, welding torch cleaning, etc.). The tool stands or the robotic cleaning stations tip dressers for welding torches should be arranged in a way that there is no need to secure them with anti-collision zones. The cell production system stations are particularly demanding in terms of protecting against collision. The sequence of robot movements should be selected in a way minimising the robot downtime. If this cannot be achieved, you may consider changing the distribution of tasks between robots.

If the robot executes different programs (e.g. welds parts for different models), each program must be checked and thought over in terms of collision avoidance. The level of difficulty increases in the case of mixed production. Some parts may be of different lengths and widths, which means that a collision can occur in the event of moving from manufacturing of workpiece A to the workpiece B. It may also be necessary to change the tool during the process, which should also be taken into account and checked.

6.2. Cycle time

Cycle time is the key parameter when planning robotic lines and writing programs for robots. It is the time of all processes that are carried out in a robotic station. It is calculated from the moment of loading a part until it is handed over to the next station (following the completion of all activities). The shorter the cycle time, the more parts we will be able to produce within a given amount of time.

The cycle time consists of the time devoted to carrying out operations by the employees (e.g. loading parts), operations performed by the robot and the time when parts are transported within the station.

Shortening of cycle time is a very unrewarding task. The effort that one needs to make in order to shorten the cycle time by the next one tenth of a second grows exponentially. The following things help to achieve the planned cycle time: the selection of the right robot, proper planning of the cells layout, optimisation of robot programs and tool design.

However, above-all, it is important to consider where accuracy could be sacrificed for higher performance. This certainly applies to the majority of movements required just to reach the first process point. Movement of a robot with a more curvy shape will be smoother and more natural. An exception to this is picking and placing, where the velocity needs to be reduced and more precision of movement is necessary. One should use the fastest axes of this particular manipulator and try to avoid rapid acceleration and braking. It would be perfect if the robot's starting position was in the same orientation as the first process point, so that robot movement will take as little time as possible.

It is necessary to think over the communication between the robot and the PLC, as a lot of time can be saved by optimising the exchange of information and minimising the time needed for data exchange (communication lag). A signal to open clamps on station, from which the part is to be taken, can be sent upon robot's movement to pick. Checking if the open condition is fulfilled should be in the closest collision-free robot position in front of the station. In this way, the robot will not wait unnecessarily, as the clamps on the fixture will open during the robot's movement, which will save some time. However, in case of breakdown station being closed, robot stops in front of the station to avoid damaging gripper and the station.

It is important to determine the correct payload mounted on the robot; an improperly declared payload affects not only the precision and durability of the robot, but also the velocity at which it moves. The right parameters allow the robot to brake or accelerate in advance in order to achieve the optimal motion trajectory.

During the design phase, it is important to check carefully whether a single tool is universal. It is possible that one may be able to design a gripper that can hold two different parts or a welding torch that can weld three different car body models. This will avoid wasting time for changing the tools.

Reducing the weight of the tool can also have a positive effect on the cycle. A lighter tool can be mounted on a robot with a lower payload capacity, which is, in general, much quicker than its larger brothers.

The cycle time is estimated at the initial design phase of a robotised station. By knowing the preliminary layout of the stations and the tasks that robots are going to perform, you can calculate the approximate time needed to execute all tasks. The next stage of validation takes place in the simulation program, where you can already see exactly how much time is required for the robot to move. These results may differ significantly from the first estimates, depending on the complexity of the trajectory. If a welding points difficult to reach, it will take much longer for the robot to reach it. At this stage you can also see which robots can work simultaneously and whether or not they block one another. If there is no problem with the cycle time at the simulation stage (in which all times have been correctly defined, e.g. gripper closure, welding gun closure and time for communication), it is highly possible that there will not be any problems on the production line as well.

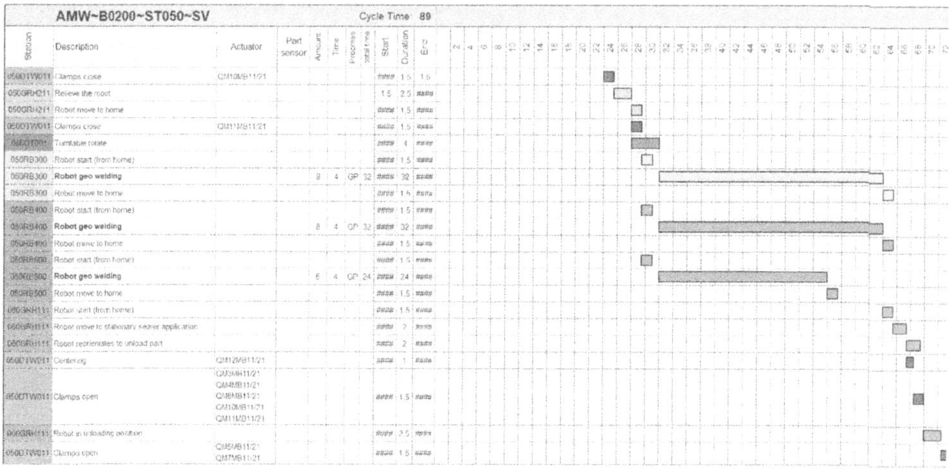

Fig. 74: Gantt chart.

The cycle time is shown in the form of Gantt charts (Fig. 74), presenting all operations together with their duration.

7

WORKING SAFELY WITH ROBOTS

Before the first industrial manipulator has been created, Isaac Asimov began to wonder what rules should govern the coexistence of humans and robots. In 1942, he formulated the famous three laws of robotics and published them in his short story titled Runaround [20]:

- A robot may not injure a human being or, through inaction, allow a human being to come to harm.
- A robot must obey the orders given it by human beings except where such orders would conflict with the First Law.
- A robot must protect its own existence as long as such protection does not conflict with the First or Second Laws.

It should be remembered that these laws were formulated for fictional devices with intelligence and human-like consciousness. The first robots known from novels and films were not shown as mechanised helpers, but as hostile invaders trying to exterminate humanity at all costs. The fear caused by this representation of robots can be seen in the later zeroth law of robotics:

> *"A robot may not harm humanity, or, by inaction, allow humanity to come to harm."*

The above rules may seem a bit funny, as there is no indication that there will be a rebellion of machines coming in the near future. However, it must be remembered that even though almost 80 years have passed since Asimov wrote these postulates, they have never been obeyed in industrial robotics. A robot can cause harm to a human being at all times, starting from minor injuries and ending with death. This is caused by the technical constraints, economic reasons and the ways organisation work on robot lines.

The Guinness Book of Records states that Robert Williams was the first person who died in an accident caused by a robot. He died on 25 January 1979 while working at the Ford foundry in Michigan. There is not much information about the course of this unfortunate event except that the robot's arm crushed the employee's head.

However, we know much more about the tragedy that Kenji Urada, the most known victim killed by a robot, has experienced. This 37-year-old maintenance engineer was working at the Kawasaki factory in Akashi, Japan. At the moment of the accident, he was close to a malfunctioning robot, which he did not turn off properly because he was in a hurry. During the repair, the robot set off and pushed the Japanese man into a grinder, which was used to process gears for car transmissions. This tragic event took place on 4 July 1981 [21].

Nowadays, fatal accidents still occur. The last famous case was the death of a worker at the Volkswagen factory in Baunatal in July 2015. A 22-year-old employee was in the middle of installing the equipment on the robot, while the robot started its work and crushed his chest. The worker died in hospital as a result of injuries he sustained in the accident. According to the data collected by the U.S. government, 33 fatal accidents caused by industrial robots have occurred in the last 30 years in the United States.

A lot of time has passed since the first accident mentioned here happened. Since then, the robot manufacturers have devoted a lot of attention to issues related to improving safety. The advances made in sensors both internal and external, have developed significantly. By monitoring the electricity consumed by electric drives, the additional resistances (and thus the possible collisions) can be easily detected. Modern robots know much more about their surroundings. However, when looking at the reports of the U.S. OSHA (*Occupational Safety and Health Administration*), one can easily see that the fatal accidents caused by robots and the related amputations are not something that took place only in the dark beginnings of the robotics development.

If you have never been in direct contact with industrial robots, you might be surprised how big they are in reality. The robot does not look dangerous on photos or computer screen. In fact, a manipulator has a payload capacity of 500 kg, it is about 3 m high, and when placed on a two-metre high platform with a mounted gripper that can lift the whole sides of the car, it looks impressive even when it is not moving. When we add to this the ability to move at a velocity of a few meters per second it starts to look really dangerous.

All robot-related works should only be carried out by an appropriately trained personnel. The absolute minimum in terms of health and safety is the basic personal protective equipment: a hard-hat and boots with an impact-resistant toe cap. Oftentimes glasses, gloves, protective clothing with long pants/sleeves, and a reflective vest or earplugs are also required. In the event of an accident, please remember that the most important thing is to protect people's life and health at all times. The safety of the machines and equipment is a secondary matter. Machines, unlike humans, can be restored to full functionality. Of course, one cannot be indifferent about the damage done to the equipment, but your own and other people's safety should always be the priority.

Serious accidents and fatalities occur every year in robotics. They are caused by: faulty equipment, not knowing how the robots work and how they are programmed, or simple human omissions caused by routine or fatigue. That is the reason why it is essential to strictly adhere to all legal standards and safety recommendations. Programming robots may pose other threats than those caused by the robots. Pay attention to all instructions for working at height or in hazardous areas (e.g. paint shops, welding shops). In addition to the hazards caused directly by the manipulator arm, there are a number of hazardous factors that may occur in robotic processes. In case of joining materials it can be radiation (laser welding), harmful gases (arc welding). Welding, in turn, means high currents and strong electromagnetic field, which makes it forbidden for people with pacemakers to work in such environments. Electronics (smartphones) also do not tolerate such conditions. During palletising, the item may fall off (due to gripper failure or human error).

Robots are usually powered by 400V, which means that any damage to cables should be reported to the appropriate people. The servomotors can heat up to high temperatures during operation and touching them may result in burns. The respective stickers on the drives (Fig. 75) warn of high temperatures. It is also not allowed to climb on the robot or carry animals or people with it (unless it is a specially adapted model offered by the manufacturer, such as the KUKA robocoaster). Yes, all those funny videos on YouTube where a robot with a mounted stool acts as a rollercoaster or a surfboard were made with a gross violation of the safety rules.

Fig. 75: Sticker – burn hazard warning.

The robot must be fastened to the ground with the proper screws, and the floor has to meet the strength requirements specified in the robot manual. It indicates the forces that may act on the floor during the robot's movement and emergency stop. Tightening the robot with sufficient force prevents it from pulling itself out of the ground. Improper installation can cause the robot to break off the screws and tip over during a dynamic movement with a heavy load.

As I mentioned in the previous chapters, a robot can be run in three modes, two test modes (T1/T2) and one automatic (Auto).

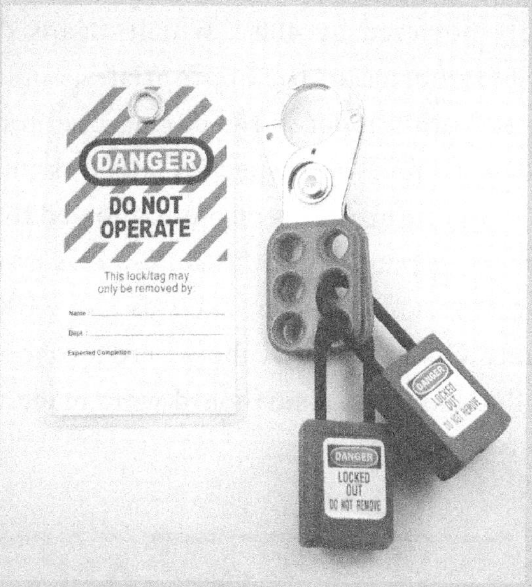

If the robotic station is equipped with a door, a lockout padlock should be used when entering the fenced area. By hanging and padlocking the door, it is not possible to close them, and a properly configured safety system will never allow the robot to operate automatically when the fencing around the robot is open. Sometimes, after closing the padlock, an employee forgets to take it with him/her after finishing work. That is why it is a good practice to put your contact details (name, company name and telephone number) on the padlock. By calling this number, a person can confirm that there is no employee inside the robot station and safely remove the padlock when it is necessary to start production.

In the first test mode (T1), which is used for teaching and testing the programs that have been written, the maximum robot velocity is limited to 250 mm/s. It is permittable to be present in the robot's work area, but no other people except for the programmer should be there. If there is a necessity for more people to be present, they should have an access to the emergency stop push buttons (e-stop), and each one of them must be able to see the robot and other people in the cell. The programmer shall be standing in a place that allows for an easy evacuation. One should avoid dangerous situations (such as standing under a robot that holds a part in the gripper).

The second test mode (T2) is designed exclusively for testing programs and allows you to run them at the full velocity declared in the program. In this mode, it is not possible to teach the robot and stay in its work area. It is also not allowed for other people to stay inside the station and the operator should stay behind the safety fence. In this mode, the tooling that the robot works with, usually operates automatically. This means that when running the program in this mode, even at low speed (e.g. within the range of 1-10% of the robot's capabilities), we are exposed to the risks associated with the execution of this process. The robot will weld (sparks!), the installed laser snout will emit a high-power beam, the spray nozzle will spray paint, etc.

In automatic mode, all safety devices must be active. The door to the station must be closed and nobody is allowed to stay inside. It is important to ensure that the robot and all secondary machines are properly protected. In the factory's accident reports, there were already records of events where the machine crushed an employee following the accidental covering of the light sensor, which resulted in sending a signal to start the machine.

For safety reasons, robot's teach pendant are equipped with three-position safety switch (dead man switch) that allows for emergency stopping. His principle of operation and functions are described in chapter 3.3 which is devoted to the construction of industrial robots.

According to EN ISO 13850:2008, there are three ways of stopping the manipulator's work.

Stop category 2: Controlled braking with leaving the connected power supply to the servomotor. The robot gently slows down until it stops completely. After stopping, the servomotors are still supplied with power, and the program execution is paused. This allows for a quick resumption of operation and does not shorten the lifespan of the servomotors. The robot stops without deviating from the motions path. Safety-related control system monitors whether the manipulator stays at the stop position.

Stop category 1: Controlled braking by disconnecting the power supply from the servomotors, after they have been stopped. The manipulator slows down until it stops completely. The operation of the servomotors allows for braking. When this happens, the brakes are applied and the servomotors are disconnected from the power supply.

Stop category 0: Uncontrolled braking, by immediate disconnection of the power supply from the drives. This type of operation interruption takes place when the emergency stop button (*e-stop*) is pushed. When the stop signal is received, the servomotors are immediately disconnected from the power supply and the brakes are applied simultaneously. The priority is to stop the manipulator as soon as possible, therefore the robot trajectory is not maintaining this method of braking can lead to a shortening of robot's lifespan due to significant overloads acting on the robot's mechanisms. In case of life or health danger, the category "0" signal should always be issued. This is the fastest way to stop the robot. The resume of work requires driving to one of the program points manually, so that the robot can return to its trajectory and start from a point with known coordinates.

Unfortunately, pushing the emergency button will not stop the robot immediately. For example, the ABB IRB 6640-205/2.75 with a range of 2.75 metres and a payload capacity of 205 kilograms, when operating in automatic mode and moving at a maximum velocity and a maximum load, needs about one second to stop (Tab. 16).

Tab. 16: Braking time and distance for ABB IRB 6640-205/2.75.

Type of braking	Robot axis	Stopping time (sec.)	Distance covered (deg.)
Category 0	Axis 1	0.9	47°
	Axis 2	0.5	22°
	Axis 3	0.3	17°
Category 1	Axis 1	1.2	60°
	Axis 2	0.9	38°
	Axis 3	0.5	24°

Stopping distance is the angle of rotation of the robot axis covered between the reception of the STOP signal and bringing the robot to a complete stop. The time needed to stop the robot completely is the stopping time.

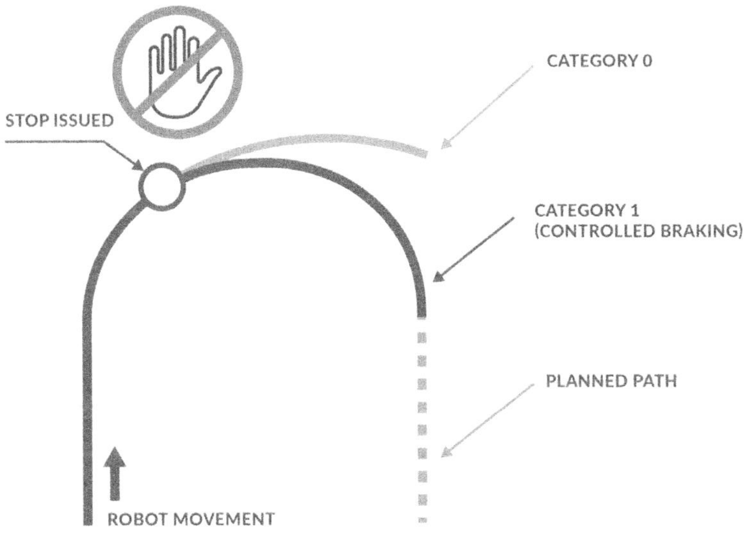

Fig. 76: Category 0 stop is the fastest, but it does not allow for the robot trajectory to be maintained.

We can distinguish several basic types of injuries that may be caused by an industrial robot. The first type of an injury is when the robot hits a person that is standing up freely. The second is a collision with a person that is partially supported (e.g. a person that is sitting). The third is hitting a completely "constrained" person (e.g. pushed to the wall or to a machine). The last type of injury is the jamming of the body parts between the robot arms. It is also possible that the injury is caused by a secondary hit that happened as a result of a direct collision (e.g. an employee falls over because of a collision with a robot and then hits the floor with his/her head).

In case of heavy industrial robots, it is not the impact energy that is the greatest threat, but the braking distance. When the robot hits a free-standing person (with an arm or a dull large-diameter tool), it is usually not a major life threat. Tab. 17 shows the speeds achieved by a person during various sports activities. You can find numerous videos on YouTube where running people suddenly bump into a post or a wall. Their speed is much higher than the walking speed shown in the table, but they are still far from braking the 100 m run world record. In most cases, such an unpleasant collision ends in a fall, but you cannot see it result in life-threatening injuries.

Tab. 17: The speeds achieved by a person during various activities.

Type of movement	Velocity (m/s)
Slow walk	0.5
Fast walk	2.0
20 km running race (men's world record 2015)	5.00
100 m running race (men's world record in 2009)	10.438

Medium and high payload capacity manipulators offer maximum linear speeds of about 2-7 m/s. Since a running person hitting a post or a wall at a speed ranging from 2 to 5 m/s is not majorly injured, similarly, a robot hitting a free-standing person at such speed should not result in a life-threatening hazard. The speed of collaborative robots is limited to just 0.25 m/s.

The studies show that when a robot hits against a free-standing person, the accident should not end tragically, but the situation changes drastically when the person stands up against a wall or a machine. By looking at the braking distance presented in Tab. 16, it is clear that even if the robot starts braking immediately when it comes in touch with a human being, it will stop only when it comes in contact with the wall, crushing everything along the way. Just to remind you, the ABB IRB 6640-205/2.75, which is commonly used in the car body welding industry, weighs 1320 kg (without a tool that can weigh up to 200 kg). A collision with a robot of this size and a man pushed against a wall, who cannot escape, could result in tragic consequences.

7.1. Collaborative robots

Over half a century ago, science-fiction writers predicted cooperation and interaction between robots and people in our daily lives. However, it was only recently that the progress made in control, construction, movement planning and safety systems made it possible to finally think about close physical Human-Robot Interaction – pHRI. In the near future, robots will help us in the course of production tasks (manipulation, assembly, simple processing), and in the long run, there will probably be personal assistants that will take care of sick and elderly people. It is still difficult to predict what form they will take, but it can be expected that the robots will soon be let inside our homes.

The emergence of a new product segment, the collaborative robots, required considering the basic safety issues. Large robots are designed to work in an area where there are no people, which is the reason why nobody has ever tried to investigate the dangers that they can cause in the event of a collision with a human being, since nothing like this should ever take place. The situation with the collaborative robots is much different, because people's presence in close proximity of the robot is desirable in this case. It was therefore necessary to develop criteria and measurement methods that would reliably assess the risk level posed by the robots.

One question still remains unanswered – what kind of injuries a robot can cause? We still have not finished identifying the critical values above which the human life and health may be threatened. The first crash tests were carried out with the use of crash-test dummies. Initially, the focus was on head, neck and chest injuries, which were identified as the most probable and most threatening to human life and health.

In the automotive industry there are standards defined that verify the safety of cars introduced to the market (Euro NCAP, NHTSA, ANCAP). They have a long tradition and their reliability has been verified by examining the actual effects of the collisions. Crash tests help determine the level of safety of the next generation of cars that are being launched on the market and compare them with one another. It would have been a very valuable piece of information if there were such standards for collaborative robots. Since there were no methodologies designed for testing of the industrial robots, the scientists had to use verified research methods from the automotive world. It should be borne in mind that the maximum speed of robots is definitely lower than that of passenger cars. In the Euro NCAP test, a frontal crash test is performed at 64 km/h (approx. 17.8 m/s). This speed is way beyond the reach of industrial robots. Not only test dummies were borrowed by the researchers, but also the classification of potential injuries and how threatening they are to human life.

DLR (*Deutsches Zentrum für Luft- und Raumfahrt*, a German space agency), together with ADAC (*Allgemeiner Deutscher Automobil-Club*, a German automobile club), conducted crash tests of robots with Hybrid III Dummy dummies (used in car safety tests) [22]. In the course of the tests, the forces applied to the head, neck and chest during crash were checked for various cases (free, restrained and partially restrained hit). It turned out that hitting the robot's arm on a freely standing person in most cases does not pose a great danger and leads to minor injuries. Positive results collected during this research allowed for a commencement of work on collaborative robots, both in terms of their construction and the law regulating their existence.

Past safety standards, intended for high performance robots ordered fencing, which limited the possibility of using them at small companies. Legislators agreed to loosen those restrictions in 2006 by drawing up ISO 10218-1:2006 standard, which officially established the category of collaborative robots. Since this type of robots was only at the stage of research back then and there were only prototypes, the regulations that were made were rather restrictive (maximum speed of 0.25 m/s, force limit: 80N and/or 150W power).

It was quite similar to the situation when the first cars started to appear. The famous *British Locomotives on Highways Act* of 1865 (better known as the "Red Flag Act") limited the speed of cars to 3.2 km/h in the city and 6.4 km/h in the country. In addition to the speed limit, a man carrying a red flag had to run in front of a road vehicle to warn passers-by about the danger.

As in the case of vehicles, it quickly turned out that the first limits set for robotics were indeed too strict. The amendment to the standard from 2011 allows for the use of higher speeds, power and pressure force. Unfortunately, there are no indications as to their value. The only thing to be considered is whether or not they are safe. This is determined by the purpose for which the robot was designed and the process that it performs. Collaborative robots which move on autonomous platforms, the so-called AMR (*Autonomous Mobile Robots*), are the next challenge for the legislative authority.

The standards define four methods of man and robot collaboration [7]:

- controlled stop (the robot stops and does not move when a human being is present in the work area, after the operator leaves the work area, the robot can continue to execute tasks at full speed) – it is possible to load or receive parts manually through the use of category two stop
- manual robot guidance (the robot moves at limited speed following an explicit request of the operator) – the manipulator serves as a supportive device, e.g. for handling of heavy elements
- distance and speed control (the robot slows down its movements to safe speed when the operator comes closer) – e.g. replacement of container with parts without interrupting the manipulator's work, as long as a safe distance is maintained
- limiting the power and movement forces of the robot (allows for an injury-free contact between the human being and the robot) – e.g. an employee can install a part working "shoulder to shoulder" with the robot.

The first three types of collaboration are now available for normal high-payload capacity robots, whereas the last category is devoted for collaborative robots with a special construction.

The main idea behind the design of collaborative robots is that they must not cause pain or injury in case they come in contact with humans. This is a quite general statement, which is difficult to translate into certain values. Apart from the individual sensitivity to pain that each one of us has, there are also problems such as the frequency of contact with the robot. While a single hit may not be perceived as a nuisance, repeated impacts may become a problem. It is unacceptable for an employee to show up at home every single day with a new bruise.

The researchers from the Mainz University have distinguished 29 points on human body and examined their pain sensitivity. 100 people participated in the research, which made it possible to make a total of

9000 measurements. On the basis of the collected data, the researchers have drawn up a table with the information about pain sensitivity which depended on the pressure force and the place of impact. All the gathered data allowed to determine the boundary values for collaborative robots. For example, if the manipulator is able to hit a man on the temple, then the pressure force must not exceed 130 N/cm2, if it is the hand that is exposed to danger, then this limit value rises to 260 N/cm2 respectively [23]. The limit force also varies according to the way of contact. If it is a short free collision (no part of the body gets crushed or jammed) and the operator can withdraw, then higher values are allowed.

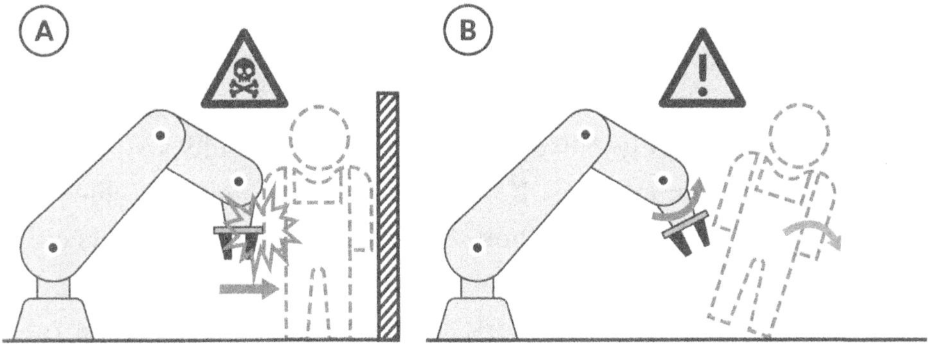

Fig. 77: Types of contact: a) quasi-static contact, b) transient contact.

These findings stood as the basis for ISO/TS 15066 document, which provides precise data that helps determine the acceptable robot performance in terms of safety. ISO/TS 15066 has not become a standard yet, but it has the status of a technical specification (TS), which means that it is still a draft version, but it is close to being published as a fully-fledged norm.

The cobots themselves are not a threat to human beings if they work in accordance with restrictive legislative regulations. The values of acting forces are adjusted in a way that makes it impossible to cause permanent injuries. It should be borne in mind that choosing a collaborative robot over other types of robots does not make the station or process safe and accessible to the employee. This rule only applies to safe processes where the robot does not have any dangerous tool installed. If the robot applies coatings or cuts materials then the same safeguards as for "big" robotics must be applied.

If the manipulator is to juggle with knives, then regardless of the level of its technical sophistication, it will always be classified as dangerous and it will be necessary to use standard safety mechanisms such as fences and separation of the machine from people.

A robot itself cannot harm the body in a way that would cause cut or lacerated wounds. A small cut can be regarded as a minor injury (from which one could protect oneself by using, for example, appropriate clothing). Please note that some vital organs, such as the liver or heart, are relatively close to the skin. One should also consider very delicate areas, i.e. eyes. If, as a result of the collision, bleeding, fracture or any damage to the bone system is possible, then it is obligatory to use fencing (regardless of the robot construction). Therefore, the construction of an open work station with a collaborative robot must be preceded by an appropriate analysis of all potential threats to employees.

In Fig. 78 you can see a decision tree, which can be used to determine the risks in an already designed robotic station. According to EN ISO 13849-1, risk is determined on the basis of three factors, and there are two options for each of them. There is no specified boundary in the standard, but the following interpretations have been adopted:

- the degree of injury (S, severity):
  - S1: mild – swelling, abrasions, stab wounds and slight crushing
  - S2: serious – bone injuries, amputations and death
- the frequency of exposure to risk (F, frequency):
  - F1: seldom to occasionally, short exposure time
  - F2: often to continuously, long exposure time

▸ the possibility of avoiding danger or limiting injuries (P, possibility):
  ◦ P1: possible under certain conditions
  ◦ P2: almost impossible

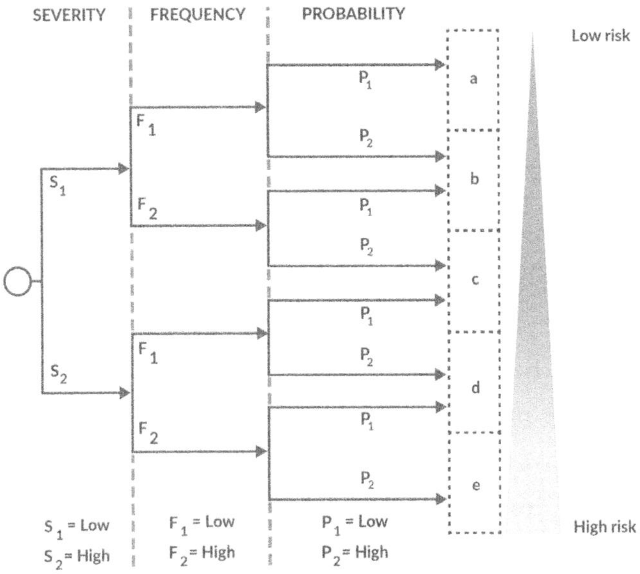

Fig. 78: Decision tree for risk assessment.

In addition to safety as measured in terms of the severity of injuries that occur in the event of a collision, the problems related to the psychological well-being of an employee becomes increasingly important. The manipulator should not take “dangerous” looking positions or make sudden movements. All movements should be smooth and predictable so as not to create unnecessary sense of danger.

In case of human-human contacts, there is continuous exchange of non-verbal signals, which allows us to determine the emotional state, level of concentration and intentions of a co-worker on an ongoing basis. We register many of these signals subconsciously without realising it (body language – looks, facial expressions, gestures). It is not possible to establish this kind of communication between a robot and a human being. What is the topic of research now is how a person reacts to a robot movement in his or her environment. This can be checked by measuring skin conductivity (skin-galvanic reaction), cardiac monitoring (electrocardiography), pupil reactions, muscle diagnostics (electromyography).

Observation of these parameters allows for determining which types of movements are most surprising or cause the person to feel anxious. This also increases the productivity of the robot. If a person does feel anxious, is focused and registers the robot's behaviour, the robot may be allowed to increase its speed. If human reactions indicate fatigue or anxiety, it would be sensible to reduce the speed of the manipulator, as there is a higher risk of an accident caused by fatigue or fear. This also relieves the operator, as some tasks can be started without any confirmation from the operator. If the robot spots that the human being gives silent acceptance, then it can proceed with the execution of the subsequent tasks. Non-verbal information, such as psychophysical signals, is usually preceded by the verbal signals from the operator. This helps to improve not only the total safety, but also improve the perceived level of safety by the employees.

The results of all these studies will be used in the future to create friendly trajectory planners, which will contribute to the increase of the level of OHS of work with and industrial robot. People usually show signs of discomfort when the robot is on their way, it is behind them or is directly on a trajectory of a collision with a human being. Robot manufacturers are also considering giving their products a human face, by mounting head models capable of sending non-verbal signals, or by using monitors that display an avatar that would be a visual representation of the robot's condition.

7.2. Safety measures

Industrial robot, as a stand-alone machine, in accordance with the EC Machinery Directive (2006/42/EC) is an unfinished machine and does not bear the CE mark. To get the certification, one needs to install proper fencing and safety systems. You should remember about this in case you want to install an industrial robot on your own.

A welding robot weighs over a ton and moves at a speed of a few metres per second, posing a fatal threat to a person who would show up in its work area. The robot controller allows you to connect external safety devices such as safety gates or light curtains. This allows the robot's safeguards to be activated both by external devices and by the robot itself.

In addition, the robot could be equipped with a collision-detection option (such as ABB MotionSupervision, KUKA Torque Monitor) which in case of detecting an increased resistance simply stops the robot. Their operation is based on monitoring the speed of the servomotor, the forces acting on it, and the electric current. In Fig. 79, you can see how these parameters change during a collision on the example of the ABB Motion Supervision system.

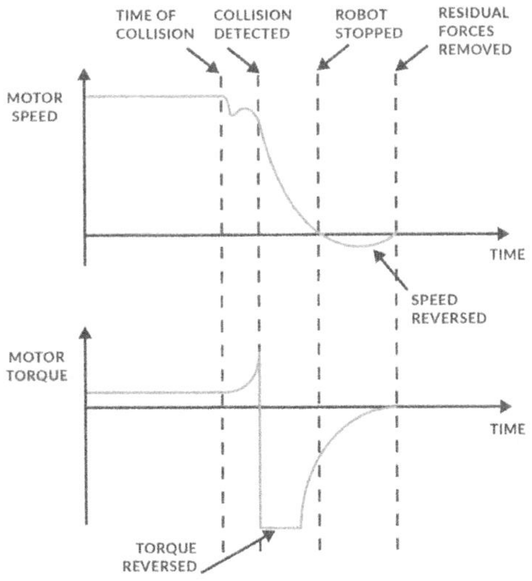

Fig. 79: ABB Motion Supervision – principle of operation.

When a collision occurs, the rotational speed drops and the torque on the motor increases. When the torque value rises above a certain level, the robot controller treats this incident as a collision. In order to minimise losses and possible threats, the robot stops and moves back as little as possible along the existing path as if it was to move away from the collision spot. The robot remains active the whole time and after the error is confirmed, it is possible to resume the execution of the program. The main aim of this type of collision detection is to protect the equipment of the robotic station. The collision detection is not sensitive enough to be considered a method that increases human safety.

The movement of each axis can be secured by imposing software limits that would restrain the manipulator's workspace. Modern robot controllers are equipped with a wide range of software options that perform safety tasks (Fanuc DCS, ABB SafeMove, KUKA Safe Robot, Yaskawa FSU). You enter tool size information and create robot work zone areas. The controller checks, in real time, the position of robot arms along with the tool installed on it, and verifies if its movement can be permitted within the defined work zone. If any safety conditions are violated (e.g. a laser scanner has detected human presence in the robot's work area), the robot will be stopped.

The work area of the industrial manipulator is guarded with fences. The entrance to this area is secured with door and proper locks (with sensors detecting when the door is open) or roll-up doors.

If the process requires the presence of a human being in the robot's work area (e.g. loading of parts), further elements are added in order to identify a violation of safety zones or the presence of the operator. These can be light curtains, pressure-sensitive safety mats, laser scanners. There are also additional safety buttons located on the fences allowing you to stop the robot without having to reach the nearest controller.

Remember that you are also responsible for your own safety. When working with the robot you should wear a protective helmet and safety shoes. These two are the minimum personal protective equipment that should be worn even if you are not a robot programmer. During maintenance work, it is very easy to smash your head against a tool mounted on the robot or against other elements in the station. Some factories require you to wear reflective vests and glasses. A failure to comply with the plant's standards can even result in firing an employee from the plant without the possibility to go back to work.

The American National Institute for Occupational Safety and Health (NIOSH) provides a list of recommendations that should be followed when constructing robotic work stations:

Use of additional sensors (motion detectors, light curtains, pressure-sensitive safety mats), which will detect human beings when they cross the fencing and they will stop the production.

Use of remote robot diagnostics tools so that the process parameters can be read without having to be in the production area. In case it is necessary to work in close-proximity to a working robot, a second employee should watch over the first one, so that he or she can stop the production in case of danger of his/her life.

All buttons (especially the safety ones) should be named clearly. Each station should have documentation showing the work area of the robot and describing the most important safety issues.

Physical fencing should meet appropriate standards and must effectively protect people from access to the hazardous area. The fences and their fastening, must be strong enough to withstand any forces that may act on them from both inside and outside. They should be installed in a way that maintains the minimum distance required from the danger zone. The locks on the door should be connected to the safety circuit (in case somebody opens the door, the production should stop).

It would be best not to approach the robot at all. And if there is the need to approach it, the robot should be stopped. Unfortunately, in case of programming an industrial robot, none of these two conditions cannot be logically met. Therefore, we should follow the following recommendations:

- Before starting to work, take a look at the work area. The work area of the robot should be clean, without any oil, water or any rubbish around.
- Before programming the robot, make sure that the control panel works – in particular the three-position safety switch and the emergency stop button (e-stop).
- When you enter the robotic station, leave the door to the station open and secure it to prevent it from closing by putting a closed padlock on the lock. If you forget about this, other employees may not notice your presence, close the door and run the station in automatic mode.
- Make sure that there is an escape route and that it is free from any obstacles.
- Each new program should be carefully checked, starting with low speeds, which should be increased in subsequent test cycles.
- If you run the program after a longer break, make sure that there are no mechanical changes to the station. A moved fence or a rebuilt gripper may cause a serious collision. All mechanical modifications shall be visibly marked (warning tapes, sheets of paper with easy-to-see information about changes).
- Before starting production in automatic mode, you must be sure that nobody is present in the robot's work area.

Special attention should be paid to the training of personnel working in robots' vicinity. All employees should know how the layout of the robotic cells and the tools that are used in the process. All persons who take part in the programming, operation and servicing of robots should be provided with an appropriate OHS training. Experienced staff should be retrained every now and then, so that they can be provided with information on the changes in procedures caused by technological developments. Employees should be aware of the risks of working with industrial robots. They should know the robot's range, the ways of safety systems operation, the location of these systems and the location of emergency stop buttons.

8

CONCLUSION

"Robots are changing the world and driving it in directions we cannot clearly foresee." – said Isaac Asimov in the foreword to his stories in 1982, and his words are even more relevant now than ever.

The industrial robots were kept hidden inside the dark corners of production halls for many years. A few years ago only few people believed that they would ever leave these places. To be honest, looking at their form back then, nobody needed them outside the factories. Large and clumsy machines could only be operated by specialists, who knew how to start and program them and, therefore, were not interesting for people outside of heavy industry. These were not the only obstacles for the machines to finally leave the plants: there was no appropriate legislation, the construction of robots at that time could not ensure appropriate safety, and there were different requirements as to our work places.

These heavy constructions and machines which are difficult to program can now be considered a thing of the past. Industrial robotics adopts all the user-friendly solutions we know from consumer products. Touch interfaces, 3D simulation software and virtual reality are the solutions that are already in everyday use. Robots can be permanently connected to the Internet, they can diagnose themselves and inform us remotely about their problems (predictive maintenance).

The legal situation governing robots and the safety associated with them is still evolving. Presently we can benefit from all the advantages of collaborative robots. Industrial manipulators are beginning to appear in places that still seem a bit too exotic for them. An example of this is the film industry, which adopts robots at a very fast pace to help with the shooting of new films. On the one hand, they allow stuntmen to make great advances, and on the other hand, they are used to manipulate the camera and shoot dynamic shots. There is also quite a lot of information about robotics in the context of agriculture and construction. Today it is still difficult to estimate how much robotisation will be in our daily life. But we know it will surely be part of it.

I strongly believe that the ability to program robots will soon be a competence that many of us will have to master. Today there is no need to be involved in the production at foundries or car factories to meet a robot. Industrial manipulators are starting to appear at trade fairs and display windows. They will accompany us more and more often in our daily tasks at work, and over time they will probably be easy to find inside our houses as well. The spread of robots in our environment resembles the future spread of autonomous cars. Nobody wonders whether it will be possible to realise, the question is only when will it happen. Robot manufacturers are in the swim and constantly improve their products, making them safer and easier to use.

Despite the enormity of the changes that are currently taking place, the most interesting times are still ahead of us. We have already seen the Autonomous Mobile Robots (AMR), i.e. collaborative manipulators which move on movable platforms, coming into existence. There is still a lot to be done in the area of safety (both in terms of construction and legal aspects). Virtual commissioning still does not cover all the scenarios that occur on the real production line. Virtual and augmented reality is still used to a limited extent. What will become of an even greater importance is cyber security – robots are now being connected not only to industrial networks and companies' intranets. More and more often they start to communicate with the outside world. Robotics follows the same principles as classic programming: there is a need to implement more and more artificial intelligence. Robots are very good at manipulating objects, but their cognitive abilities are still very limited. There is a strong demand for advanced algorithms that process and classify the recorded images. As you can see, there are still many areas in robotics that can be improved.

At the same time, we can see that the labour market is changing as well. Not many of us want to perform hard, physical work. We've always wanted to have a mechanical helper that would relieve us from repetitive and strenuous tasks. It is possible now, but we have to get ourselves acquainted with the ways of operating it. You do not need to have an access to a real robot to get to know the rules of robotics. Today, most things can be tested in virtual simulators.

Robotics gives an opportunity to travel around the world, get away from your desk and work on interesting automotive (and other types of!) projects. Programming robots gives a lot of satisfaction: new things are created before our very eyes. After one day of work, you would have not only a few hundred lines of code, but also a tangible object which has been created by the robot you have controlled. You can see how a new item is made. In "ordinary" programming known from the IT industry there is a lot of code to write. The code carries out abstract tasks, which are difficult to explain to someone who is not from the industry. Robot programmer's job is much simpler – when somebody asks you about the last project you have completed, you can simply point out a car driving on the street, and tell this person about the vehicle's successor that will reach the streets in two years' time.

I hope that after reading this book you will have a chance to take part in many interesting projects from various industries. Maybe you will be one of those people who will push this relatively young industry forward. Good luck!

What's next?

Thank you for the time you have devoted to reading this book. I hope it has helped you find the answers to all your questions. Remember that apart from the content you have found in this book, I am still expanding it by putting multimedia and additional materials that can be found at roboticsbook.com/bonus/

If you would like to give me your thoughts on this book, please send me a message on Twitter at @m_gurg. You can also rate it on Amazon or Goodreads, I'm looking forward to your opinion!

9

BIBLIOGRAPHY

[1] „Ice trade", Wikipedia. 03-mar-2018.

[2] Čapek K., R.U.R. Booklassic, 2015.

[3] „Steam Horse locomotive", Wikipedia. 31-maj-2018.

[4] „Manufacturing Engineering - Masters of Manufacturing: Joseph F. Engelberger", [Online:] http://www.sme.org/MEMagazine/Article.aspx?id=27289&taxid=1434. [Dostęp: 23-cze-2018].

[5] IFR, „International Federation of Robotics - Robot History", IFR International Federation of Robotics. [Online:] https://ifr.org/robot-history. [Dostęp: 01-lip-2018].

[6] Georgakas D., i Surkin M., Detroit, I do mind dying. Cambridge, Mass.: South End Press, 1998.

[7] „ISO 10218-1:2011 Robots and robotic devices -- Safety requirements for industrial robots -- Part 1: Robots", .

[8] „Product specification - IRB 6700", 2018.

[9] „IRB 1200, M2004, Product specification", [Online:] http://search-ext.abb.com/library/Download.aspx?DocumentID=3HAC046982-001&LanguageCode=en&DocumentPartId=&Action=Launch. [Dostęp: 15-lis-2018].

[10] Calcagno R., Rusinà F., i Bonivento A., Application of wireless technologies in automotive production systems.

[11] „Data sheet IRB 6650S, PDF", [Online:] https://search-ext.abb.com/library/Download.aspx?DocumentID=PR10262EN_R6&LanguageCode=en&DocumentPartId=&Action=Launch. [Dostęp: 15-lis-2018].

[12] „IRB 6640, M2004, Product specification", [Online:] https://search-ext.abb.com/library/Download.aspx?DocumentID=3HAC028284-001&LanguageCode=en&DocumentPartId=&Action=Launch. [Dostęp: 15-lis-2018].

[13] Baroncelli A., „Comau Smart Robots Designed for Technological Applications", paź-2012.

[14] „ISO 14644-1:2015 Cleanrooms and associated controlled environments -- Part 1: Classification of air cleanliness by particle concentration", .

[15] „FAST picker TP80 4 axis industrial robot - picker", [Online:] https://www.staubli.com/en/file/5252.show. [Dostęp: 25-cze-2018].

[16] „IRB 360 Data sheet A4", [Online:] https://search-ext.abb.com/library/Download.aspx?DocumentID=ROB0082EN_G&LanguageCode=en&DocumentPartId=&Action=Launch. [Dostęp: 25-cze-2018].

[17] Daimler, „Key Figures Environmental Protection", Daimler. [Online:] https://www.daimler.com/sustainability/key-figures-environmental-protection.html. [Dostęp: 14-cze-2018].

[18] Meike D., „Increasing Energy Efficiency of Robotized Production Systems in Automobile Manufacturing", 2013.

[19] „Boy, do Fanuc and Cisco have a deal for your factory", Nikkei Asian Review. [Online:] https://asia.nikkei.com/Business/Boy-do-Fanuc-and-Cisco-have-a--deal-for-your-factory. [Dostęp: 14-cze-2018].

[20] Asimov I., i Szmigiel E., Świat robotów. Warszawa: Varia APD, 1993.

[21] „Kenji Urada - Wikipedia, the free encyclopedia", [Online:] http://en.wikipedia.org/wiki/Kenji_Urada. [Dostęp: 14-lip-2017].

[22] Haddadin S., Albu-Schäffer A., i Hirzinger G., Safe Physical Human-Robot Interaction: Measurements, Analysis & New Insights. .

[23] „ISO/TS 15066:2016 Robots and robotic devices -- Collaborative robots", .

Notes

Notes

Notes

Notes

Notes

Made in the USA
Monee, IL
20 September 2022

14316102R00122